AF522010

The Conquest of Space

From "By Rocket to the Moon," courtesy UFA Films Inc.

THE ROCKET IN FLIGHT

The Conquest of Space

by

David Lasser

with the original foreword by
Dr. H.H. Sheldon

and a new introduction by
Sir Arthur C. Clarke

An Apogee Books Publication

We acknowledge the financial support of the Government of Canada through the Book Publishing Industry Development Program for our publishing activities.

Published by Apogee Books, Box 62034, Burlington, Ontario, Canada, L7R 4K2, http://www.cgpublishing.com

Printed and bound in Canada

The Conquest of Space by David Lasser
Edited and compiled by Robert Godwin

ISBN 1-896522-92-0 - ISSN 1496-6921

Contents

TO

MAX VALIER

THE FIRST TO GIVE HIS LIFE
FOR THE CONQUEST OF SPACE

INTRODUCTION

It may be a false memory, yet it still seems as though it happened only yesterday…

The date is 1931, and the 14-year-old Arthur Clarke is walking with his beloved Aunt Nellie, past the W H Smith bookstore in Minehead only a few hundred yards from the house in which he was born. He catches sight of a book in the window…

Now, young Arthur had always been interested in astronomy, and was an avid reader of science fiction, but it had never occurred to him that space travel might one day be for real. So he persuaded Aunt Nellie to purchase the book, and it literally changed his life.

The Conquest of Space was the first serious discussion of astronautics in English. Now that men have walked on the Moon, and our robot probes have visited all the planets except Pluto, it is hard to recall the total incredulity once evoked by any talk of space travel. As late as 1956, Britain's Astronomer Royal exclaimed "Space travel is utter bilge". He never quite lived it down.

David Lasser was subjected to the same kind of ridicule, most famously when he applied for a position as a Labour Union executive. One Congressman pointed out that anyone who claimed we might fly to the Moon must be a lunatic, so Lasser was obviously unsuited for any position of responsibility.

I am happy to say that I met David once, as a very old man, when he attended one of my lectures in California. So I was able to personally thank him for his role as the first stage booster that launched me into orbit.

Arthur C Clarke
5 June 2002
Colombo, Sri Lanka

DAVID LASSER
(1902-1996)

Long before the first astronaut launched into space, visionaries foresaw the possibilities of the rocket. David Lasser was one of those visionaries. Although his involvement in the nascent spaceflight movement in the United States was brief, he left an indelible impression on future generations that would follow the dream of spaceflight.

Lasser was born on March 20, 1902 in Baltimore, Maryland to Russian emigrants Leonard and Lena Lasser. He dropped out of high school at an early age to help support his family. When the U.S. entered World War I, an underage Lasser volunteered for the American Expeditionary Forces and saw action on the battlefields of Europe. Following his discharge from service in 1919, Lasser took advantage of funds available to disabled veterans (he was gassed during the Argonne offensive) to pursue a college education. Despite the absence of a high school diploma, Lasser began studies in electrical engineering at the Newark College of Engineering, while simultaneously pursuing his high school diploma through night school. After one year at Newark, he applied for admission to MIT, from which he obtained a Bachelor of Science degree in Engineering Administration in 1924. Following graduation, he held a variety of jobs, including a position with New York Edison (from which he was fired for protesting the dismissal of several employees on the basis of time-motion study results) before landing a position in 1929 as Managing Editor of Hugo Gernsback's *Science Wonder Stories* magazine.

As Editor of *Science Wonder Stories* (which merged with *Air Wonder Stories* in 1930 under the title *Wonder Stories*), Lasser sought to raise publishing standards by encouraging writers to develop stories with more of a scientific, rather than a fantastic, basis. In addition, he took it upon himself to investigate the possibilities of using rockets to travel into space. Convinced that spaceflight was technically feasible, Lasser and several writers formed the American Interplanetary Society (AIS) in 1930, which he served as its first president. The AIS founders saw spaceflight as a means "to enlarge man's intellectual and spiritual life." To reflect growing interest amongst its members in the design and launch of rockets, the AIS changed its name to the American Rocket Society (ARS) in 1934. In 1963, the ARS merged with the Institute of Aeronautical Sciences to form the American Institute of Aeronautics and Astronautics (AIAA).

Lasser also saw cooperation in space as an opportunity for uniting mankind. In his 1931 annual report to the AIS, Lasser indicated that "the solution of interplanetary problems is too large to be localized in any group or nation." He foresaw "the building of the first space ship only as a joint effort of a united earth." Lasser joined Robert Esnault-Pelterie, the noted French astronautics pioneer, in a 1931 letter to Hermann Oberth calling for the formation of an International Commission for Astronautics. Although their call went unheeded, the concept was forerunner to the International Astronautical Federation (IAF) that formed in 1950.

As an outgrowth of his earlier investigations, Lasser wrote a popular, yet scientific, account of the history, current state, and future applications of rockets. Unable to find a publisher, Lasser and several colleagues published *The Conquest of Space*, at personal expense, under the imprint of Penguin Press (no relation to the current imprint of the same name). This was the first English-language book of speculative non-fiction to address the possibility of spaceflight (the British edition was published in 1932 by Hurst & Blackett). In this book, and in reports to the AIS, Lasser foresaw many of the commercial, scientific, and military potentials of the rocket. Although the idea of spaceflight was viewed with considerable skepticism in the science and engineering communities at that time (less than 30 years after the Wright brother's first aeroplane flight at Kitty Hawk), the book inspired more than a few people. The noted science fiction author, Arthur C. Clarke, acquired a copy as a child and later reflected that it represented "one of the turning points of [his] life."

Lasser's interest in spaceflight, however, was soon overcome by national and world events. During Lasser's employment with Gernsback, the United States was in the midst of the Great Depression. Lasser felt strongly that "unless we solved the problem of the unemployed, all else would be academic." Gernsback became unhappy with media reports of Lasser's activities on behalf of the unemployed and encouraged him to join their ranks. As a result, Lasser left his position with *Wonder Stories* and resigned from the AIS in 1933 to form a local union for the unemployed that went national in 1935 as the Workers Alliance of America (WAA) with Lasser as its president.

Although he held socialist beliefs, Lasser was strongly anti-Communist. He resigned from the WAA in 1940 due to the growing influence of Communist factions attempting to gain control of the WAA and formed a non-communist group called the American Security Union. With war raging in Europe once again, Lasser proposed to President Roosevelt the formation of a large program to provide the unemployed with the skills and training needed to support the industrial effort. Roosevelt accepted Lasser's proposal and asked him to join the program as a consultant. Although Congress accepted the President's proposal, they blocked Lasser's appointment as a special consultant with one Congressman denouncing Lasser from the floor of the House as "not only a radical but a crackpot with mental delusions that we can travel to the moon." After a period of unemployment, Lasser took a job as a labor consultant with the War Production Board, which he held for the duration of the war.

Following the end of World War II, Lasser wrote *Private Monopoly – The Enemy at Home* (NY: Harper & Bros., 1945) to publicize his views on the relationship between big business and the Federal government during the war. This did little to endear him to big business and its political influence. In 1948, W. Averell Harriman asked Lasser to join his staff in implementing the Marshall plan to rebuild Europe and counter the communist influence in the European labor movement. Although Lasser served in a temporary capacity for three months, Congress again blocked his permanent appointment due to his earlier involvement with the WAA, which the FBI had listed as a radical organization. In the face of adversity, Lasser persevered and continued to play a significant role in U.S. labor history. From 1950 until his retirement in 1969, he served as Economics and Research Director for the International Union of Electrical Radio and Machine Workers. Throughout his career, Lasser persisted in efforts to clear his name and in 1980 received a letter of apology from President Carter for being "treated unjustly."

Lasser always maintained an abiding interest in the future of humanity – this was a common theme in all of his endeavors. In a 1978 letter to Lasser, Harriman wrote, "your life represents the best in humanitarianism and in Americanism." Lasser's efforts on behalf of both the space movement and the labor movement provided inspiration and hope for many.

When Lasser witnessed the televised landing of Apollo 11 on the moon, with family and friends, he humbly indicated that his contribution to the effort was brief and not particularly vital – history will surely remember him otherwise.

Michael Ciancone

- Mr Ciancone is Chairman of the AAS History Committee, and the Executive Secretary of the Shuttle/ISS Payload Safety Review Panel at the Johnson Space Center.

FOREWORD

MAN has now reached almost every goal he has set for himself. He wanted transportation: animals sufficed for a time; but they have been superseded by devices propelled by engines. Man wanted to speak to absent friends — the wildest kind of a dream — but the telephone, and now wireless, have satisfied him. Then he wanted to see at a distance: for a time the telescope seemed marvellously to fill this need; but now we have television. He can see his friends across oceans of darkness, and he can talk to them.

And so we might go on. Wild visions have become realities. Men wanted to fly like birds; and fools risked their lives and lost them amid jeers from the crowd. But now we fly! We circle the globe in a matter of a few days. "It cannot be done" is no longer a sentence applicable to aviation.

There is but one goal left for man to attack the upper atmosphere and beyond. By means of balloons and aeroplanes we have accomplished something in this direction. But these vehicles are forever limited in the heights which they may reach by their dependence upon the air itself for lift or propulsion, or both. We can hope to accomplish little more than has already been done by these means. He may go slightly higher: but hardly enough to make the effort worth while scientifically.

But one vehicle remains to us — the rocket. Independent of the necessity of any material medium for travel it can go anywhere inside or outside of our atmosphere. It would reach its greatest efficiency in interstellar space. In fact, once started in such space it would continue, as would any moving object, without further expenditure of fuel.

Most people do not take the rocket seriously. Indeed, one can readily divide all people into two distinct classes on this matter. There are those who are satisfied that the rocket as a mode of propulsion is ridiculous, and that all rocket enthusiasts are mental defectives. These persons, almost without exception, have never given the rocket any serious thought or studied the rocket even superficially. The second division consists of those who know what a rocket is and have followed some of the recent experiments both here and abroad. They are convinced that the rocket has possibilities, but that the road to its utilization is a long one. Few are enthusiasts: their attitude is more that of the engineer who sees the difficulties ahead.

While these two classes are quite distinct, the superior attitude of the uninformed is not unexpected. One of the great contributors to the science of aviation, Robert

Esnault-Pelterie, is also a keen student of rockets. The writer recently had the opportunity of asking him directly if he was often mistaken for some kind of crank when on frequent occasions he advocated rocket study. "Not nearly so often," he said, "as when, about thirty years ago, I was experimenting with aeroplanes." Apparently some progress has been made.

There can be no doubt whatever that the number of persons interested in rocket experimentation would be enormously increased if they could but be told something of what it is all about. If up to now one were to come seeking information concerning rockets there would have been no books to which to refer him in the entire English language. A few books in Russian, a couple in German, and one in French constitute the entire collected literature. And unless the inquirer read one or more of these languages well, or was more than usually interested, he would be likely to throw up his hands and forget about it.

With the coming of the present book all this is remedied. It is the first book on rockets in the English language; it is written, as is evident, by a man with a keen appreciation of rockets and their possibilities, both for success and disappointment.

In writing this book Mr. Lasser has displayed a fine sense of balance. He has not introduced enough technical matter to discourage the poorly equipped in these matters, nor has he left out so much as to disappoint the scientist. Yet he has told all that anyone needs to know to understand the rocket principle. The scientist will automatically fill in where need be as he reads.

But Mr. Lasser has accomplished a far more difficult feat than this; he has brought into the book the enthusiasm and vision that actuates scientists and inventors. It requires much imagination to be a great scientist. But scientists the world over try, and usually succeed, in covering this up to the outside world. They write using many equations and cold sentences. You see the enthusiast only when, as a friend, you are admitted behind laboratory doors and see experiments in progress, and are told what is expected of them. They seldom fulfill expectations. Mr. Lasser has written like a scientist off his guard for the moment. That is one reason why this book is interesting even though you may detest rockets, from the Fourth of July variety upward.

Probably the most surprising thing to the uninitiated who reads this book will be the state of development which rocket study has already reached. Most people have no idea of the substantial progress that has been made by such men as Goddard in America; Esnault-Pelterie in France; Valier, Oberth, and others in Germany.

The new science is well on its way, and in several countries societies have been formed to promote the study. Mr. Lasser is a pioneer in this field, and President of the American Interplanetary Society. He has kept in close touch with rocket study both here and abroad, and is perhaps in a better position than anyone else in this country to speak with authority concerning experiments in every corner of the world. The publishing of this book will undoubtedly prove to be a milestone in the progress of rocket study in this country.

New York University H. H. SHELDON

PREFACE

AMONG unprejudiced and far-sighted men of science there is agreement today that man has in the rocket an engine to carry him away from the earth, across hundreds of thousands and millions of miles to the moon, or Mars, or Venus. And pursuing the development of that rocket, scientists in half a dozen nations are now laboring with energy and enthusiasm.

It is the purpose of this volume to present the background of the programme for the utilization of the rocket — not only as a revolutionary means of terrestrial transportation (which shall bring Europe and America within one hour's journey of each other), but also for the conquest of the planets; to point out the roads to the fulfillment of these ambitious aims, the experiences that man may encounter on the way, and what he may find upon other worlds to justify his unparalleled adventure into the unknown.

It is hoped by this presentation of the future of rocket flights in the earth's atmosphere and in interplanetary space that the mists of misunderstanding, ignorance, and prejudice that surround the "interplanetary rocket" question may be cleared up.

Much of a fragmentary nature has already appeared in the daily press and in numerous periodicals, treating of rockets and their possibilities, but no comprehensive summary for the layman, as well as for the scientist, has yet been published in English. If this work results in the creating of an intelligent appreciation of the enormous possibilities that slumber in the rocket, it will have served its purpose.

The author wishes to express his gratitude to the Scientific American, the Popular Book Corporation, The New York Herald-Tribune, and Nature Magazine for permission to utilize material that he had previously contributed to their pages. He also expresses gratitude to the Macmillan Company for permission to quote from the illuminating discourses of Sir James Jeans in The Universe Around Us, and to Doubleday, Doran and Co., for permission to quote from Professor Willem J. Luyten's Pageant of the Stars.

To the members of the American Interplanetary Society, for their co-operation, their help and their warming enthusiasm toward this volume, the author cannot offer sufficient appreciation.

DAVID LASSER.

Part I

THE ROCKET

CHAPTER I

THE MEANING OF SPACE FLIGHT

I

To the imaginative mind viewing the star-filled heavens and the luminous glow of the planets there is present continually a world of mystery, and the suggestion of tremendous adventure beyond the earth.

Since the day when man discovered the planets to be worlds similar to our earth two questions have filled his mind: How can I get there? and what will I find?

Literature from the time of the Greeks to our own era has been filled with commentaries on the nature of our sister planets and of the great interplanetary spaces. Investigators searched out the mysteries of the heavens, building up laboriously the impressive science of astronomy; and by their side were men of letters, with unleashed imaginations, peopling the heavens and the planets with the strange creatures of their fancies.

In all this investigation and romancing it was not questioned that some day the planetary mysteries would be solved. Daring men would travel to the moon on chariots propelled by the "swans of the Indies;" as Bishop Godwin, writing in the seventeenth century, stated; or in a balloon, as Poe suggested in *The Unparalleled Adventure of One Hans Pfaall*; or in the shot of Jules Verne. Exploring unknown worlds, he would discover for himself the various forms of Nature's creations.

Throughout the centuries the imagination of man has been kept alive to the supreme adventure of escaping from earth's chains and of travelling into the great spaces to land upon another world.

Some writers looked upon the interplanetary journey as a means of colonizing another planet and a method of establishing a social order freed from terrestrial evils; others were eager to explore interplanetary space and the surface of our sister worlds in the hope of epochal discoveries in science; there was an almost universal desire to find life upon the moon and Mars and Venus, and it was naturally hoped that that life might be similar to, though intriguingly different from, our own. And still other imaginative astronauts were fired by the thrill of adventure for its own sake.

The dreams of scientists and men of letters alike, however, until the dawn of the present century, were not burdened by the grim realities of the journey. It was not until the nature of interplanetary space was determined that the fanciful ideas for interplanetary travel were abandoned and a real science of spatial navigation formulated. Similarly, it was not until high-powered telescopes and the spectroscope had been trained on the moon and the other planets that any speculations about their habitability yielded to cold-blooded analyses of what life they really could support.

Fancy, then, has been displaced by science — romance has given way to the formulae of the engineer and the mathematical charts of the astronomer. But the magnitude of the task of the astronaut is still enormous. There are for him thrills never before known to man. There is depthless space to be explored, and there are unknown worlds to be trodden, strange experiences to try his skill, his courage, and his faith.

So although the interplanetary journey will be, in the last analysis, the result of the highest achievements of modern science, it will also open the way to experiences more bizarre than the dreams of a romancer.

II

Any discussion of the exploration of space, and how it may be accomplished, must rest on an understanding of the nature of the interplanetary problem. We must recognize the immensity of the cosmic forms to which we will be subject once we leave the vicinity of the earth.

The Atlantic in the time of Columbus was a sea of unknown terrors, extending to the very brink of nothingness. Yet today we sail or fly the ocean, and the cables that lie beneath it, and through the ether above, we transfer our thoughts and orders from continent to continent almost instantaneously.

We have acquired, as a result, a planetary instead of a provincial outlook — the whole earth is our home. It is but a single step from this to the acquisition of an interplanetary mind and to an extension of our concept to include the solar system.

If we can grasp standards beyond those merely terrestrial, envision millions instead of thousands of miles, and appreciate the immense but perfect harmonies of our solar system, an understanding of the conquest of space will soon follow. We will perceive then that an interplanetary journey can be achieved through the medium of the same laws of physics and chemistry that gave us the aeroplane and motor-car. We will see that a journey to the moon may be planned with the exactitude of an aeroplane flight round the earth. There is nothing, in short, of the mystic or the supernatural in our reaching out to the untrodden heavens.

To understand the magnitude of any proposed flight we must picture the earth in the heavens — a small planet in a family of planets tucked away in a corner of the universe. Whirling round the central sun, in company with eight sister worlds, from little Mercury to Pluto, on the outer fringe of the solar system, and their satellite moons and hundreds of asteroids — the earth pursues in the skies its invariable course.

Between these asserted numbers of the solar family is the emptiness of space, airless and lifeless — open for exploration. Beyond the last outpost of our system is the void, and over 24,000,000,000,000 miles must be crossed before the next star is encountered.

The sun's gravitational influence holds each of its offspring to its appointed course. The earth, with its frightening velocity through space of nearly nineteen miles per second, would shoot off to become a wanderer through the infinite did not the sun, 92,000,000 miles away, hold it fast.

With mathematical regularity, as invariable as Fate, the earth and its sisters swing ceaselessly about the great luminary, and it is this regularity, this predictability of the motions of the planets and of the natural laws that govern them, that makes possible the science of astronautics.

Through the force of gravitation, similarly, the earth binds all mankind to its surface. Though we spin on this whirling globe at almost a thousand miles an hour through the heavens, we are held to the mother in safety.

Were gravitation suddenly to vanish all things not rigidly attached to the earth would shoot violently into depthless space.

Upon the speck of dust that is the earth, therefore surrounded by the millions and billions of miles of emptiness with only the moon and distant planets as neighbors we are born, we live and die. We are the blind subjects of blind forces, pebbles upon the endless shore of the Cosmos.

III

To escape from the earth, which has chained us to her from the beginning of time, we must construct a vehicle to overcome her relentless gravitational pull; penetrate the 250 miles of air that blankets, her, and emerge into the vacuum of interplanetary space. To visit the moon requires a journey of 240,000 miles across the void, and to reach Mars or Venus a gulf of 30,000,000 to 50,000,000 miles must be traversed.

The vehicle must be able not only to lift itself away from the earth, but also have sufficient energy to continue to combat the earth's pull. Otherwise the ship would crash back to destruction on the earth's surface. For the gravitational sphere of the earth does not, pleasant romancers of former ages believed, cease once we are beyond her atmosphere. It continues, though diminishing in intensity, to infinity.

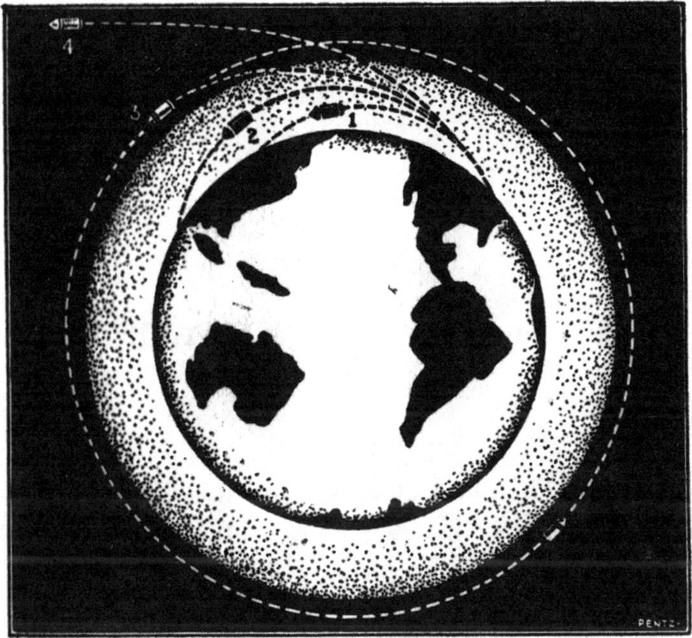

FIG. 1 THE PATH OF PROJECTILES SHOT ABOVE THE EARTH

Shots 1 and 2 will return to earth. Shot 3, fired at 5 miles a second, will circle the earth as a satellite. Shot 4, with a speed of 6.664 miles a second, will escape the earth for ever.

To be free of the earth's pull the ship must be shot away from it with a velocity of approximately seven miles per second, or a hundred times the speed of the fastest aeroplane.[1]

1 In order to understand this necessity let us assume that the interplanetary journey is to be made by a stone. We throw it upward into the air, and it returns to earth when its initial velocity has been overcome by the continued pull of the earth. If we throw it up with greater force (a larger velocity) it will rise to a greater height before its upward movement ceases. And, finally, if we can throw it hard enough (so that it attains a speed of seven miles per second) it is forever free of the earth; for as it moves away from the earth, and its speed is reduced by the earth's pull, that pull is itself reduced by the increasing distance of the stone from the earth. The speed away from the earth due to earth gravitation will never reach zero.

This speed, however, must not be attained too quickly, for in passing through the atmosphere in such meteoric fashion the ship would be heated almost to incandescence. It must start slowly from the earth, acquire velocity gradually, and reach its critical seven-mile-a-second speed only when it emerges into airless space. The continual operation of the ship's engines, even where no air exists, is thus a prime necessity.

The ship must also be capable of guidance and maneuvering in interplanetary space. As will be seen later, the space flight requires complete adherence to a rigidly plotted course through the heavens.

Since human beings are to guide the craft upon its journey, provision must be made within its walls for adequate supplies of food, water, and air, and for the maintenance of a comfortable temperature.

CHAPTER II

SPACE FLIGHT IN LITERATURE

I

It may be interesting and helpful, before going on to a discussion of how modern science may realize the space flight, to picture the imaginative conceptions of the romancers of past ages.

Although their works have not of themselves contributed much of scientific value, they served as stimulants to thought — and they aroused and kept alive the expectations of discoveries in science until such time as Science could of herself provide fulfillment. This was particularly true of the conquest of space. During the centuries that man struggled painfully to discover the nature of space and the cosmos, and to lift himself above the earth in crude machines of his own making, dozens of writers of fiction carried their characters to the moon, the sun, and the solar planets, and imaginatively explored the mysteries of the universe.

The history of the idea of interplanetary travel, therefore, is linked up quite as much with men like Cyrano de Bergerac, Poe, Verne, Wells, and Gail as it is with Copernicus and Newton, who gave us our modern knowledge of space, and Montgolfier and Wright, who provided the first devices for leaving the earth. Though men of past centuries jeered at the pictures of Lunar civilization presented by irrepressible writers, the thought remained that man had gone to the moon. When Poe wrote his *Adventure of Hans Pfaall*, and warned us against a naive belief in Hans' Lunar journey, he knew that he was none the less opening the door for a consideration of that journey's possibilities. And in modern years, with schoolboys being brought up on the novels of Verne, Wells, Gail, and others, in which interplanetary flights are considered as accomplished facts, the foundation of public support is being laid for the flight of the future.

Prior to the last fifty years the writer of fanciful interplanetary journeys claimed little special knowledge of the mechanics of his subject. He did not, in fact, take the journey too seriously. His purpose was to take his character to the moon or to Mars, and there to paint with perfect freedom a satire on our earthly civilization.

Daniel Defoe in 1700 used *The Consolidator* as a spacecraft to plunge his character into a strange lunar civilization. He is gaily naive about the mechanics of his ship, and he does not ask for belief.

> "A certain engine" it was, "formed in the shape of a chariot, on the backs of two vast bodies with extended wings which spread about 50 yards of breadth." They were composed "of feathers so nicely put together that no air could pass; and the bodies were made of Lunar earth which could bear the fire; the cavities pure filled with an ambient flame which fed on certain spirits deposited in proper quantity to last out the journey…"

There are no dangers, no terrors, on this delightfully pleasant exploration for the "person being placed in this air chariot drinks a certain dozing draught that throws him into a gentle slumber and dreaming all the way never awakens until he comes to his journey's end."

Although Defoe was vague on the method of getting his explorer to the moon, the contrary was true of what his character finds there. The moon is inhabited with human beings, we learn, who believe that the earth is their satellite, just as we believe the moon to be ours, and "they had a good deal more knowledge of things than we in this world — and nature, science, and reason had obtained great improvement in the Lunar world."

II

What confused these writers of extravaganzas throughout the ages was a lack of knowledge on which to base a description of an interplanetary vessel. It was not until the time of Newton that it was known that the earth's gravitation did not cease abruptly a few miles above the earth, but that it extended in diminishing power to an infinite distance. Neither was it known until more exact instruments of heavenly observation were invented that the earth's atmosphere extended only a few hundred miles above its surface. The two most important necessities for a spacecraft — its ability to produce a prodigious amount of power to escape the earth's gravitation, and the providing of protection against the airlessness and the extremes of temperature in space — were therefore overlooked.

Even so astute a thinker as the young John Wilkins, later Bishop of Chester, erred in both points in his serious work *The Discovery of a World in the Moone*, published in 1638. Wilkins assumed that twenty miles above the earth "weight ceased," and "if a man were above the sphere of this magnetic virtue [gravitation] which proceeds from the earth, he might stand there as firmly in the open as he can now on the ground, and he may also move about with grater swiftness; than any living creature below."

Wilkins meant his book to be taken seriously. He believed that the craft used in a moon fantasy *The Man in the Moone* — written a few years before, but published posthumously in the same year as Wilkins' book — might serve as a basis for raising us from the earth to the twenty-mile height. Bishop Godwin, who wrote The Man in the Moone, assumed that the moon was the refuge of earthly fowls for half the year. Therefore "the swans of the East Indies could be taught to carry men by having a machine to divide the weight between them." Man could then travel to the moon and return with the migrations of the birds!

Wilkins assumed falsely, too, that although the air above the earth might conceivably thin out and become colder, as it does on mountain tops, there is no actual vacuum in space, and easy provision could be made for human sustenance. For air to breathe man might use a wet handkerchief, which he presses against his nose; and for food on the way he could exist on "smells and odors"!

The ideas of Wilkins were of value chiefly because they came at a time when the declaration of Galileo that the moon was a world, like our earth, had not yet been accepted. Wilkins was not content to prove alone that the moon could be a world, but also that it might conceivably be inhabited, and that we might at some time journey there. In general terms he "does seriously affirm it possible to make a flying chariot in which a man may sit and give such a motion unto it as shall convey him through the air. It thus might be made large enough to carry divers men at the same time together with food for their viaticum and commodities for traffic. It is not the bigness of anything that can hinder its motion if the motive faculty be answerable thereto." In that final sentence Wilkins was unknowingly anticipating the objections of those who stand aghast at the plans for a giant interplanetary rocket.

" The perfecting of such an instrument," he goes on to add, "would be of such excellent use that it were enough to make a man famous and the age also in which he lives."

At the same time that Wilkins was anonymously publishing his Discovery in England Cyrano do Bergerac, a social satirist, was braving official opinion by his *Voyages to the Moon and the Sun.* Few books caused as much stir and so profoundly shocked conservative seventeenth-century France as did the adventures of Cyrano's hero in other worlds.

Cyrano, unlike Wilkins, was openly contemptuous of the ignorance and superstition of his day regarding the nature of the heavenly bodies. His explorer will go to the moon if only to prove to stupid earthlings that the moon is really a world like the earth!

Cyrano, however, is also at a loss for a vehicle to use for transportation. He has his astronaut experiment with "a belt holding bottles filled with water," that the sun will change into dew, and so draw him into the air. This device evidently works, but the explorer finds himself descending to earth again in New Canada, Cyrano sends him aloft a second time in a box-like cell, and this time it is propelled by rockets!

This was in 1640 — almost three centuries before serious scientists thought to utilize the rocket for a space flight. It is impossible to tell where this uncanny use of the only sensible means of propulsion was obtained by Cyrano. He does not seem cognizant of the fact that he chose the only vehicle that could combat gravitation and the airlessness of space. Yet he is aware of the requirements of the rocket. He sees that it can function only on a continued supply of fuel. For as the story goes on the rockets that have carried his explorer to the clouds become exhausted; the craft drops back to earth — but the explorer continues blithely on to a safe albeit undignified landing on the moon.

III

The stories of Cyrano, Defoe, Godwin, and Lucian (who wrote a moon comedy in the second century A.D.) made no attempt at verisimilitude. The writers had little knowledge of astronomy or mechanics, nor were they concerned with the plausibility of their adventures. It was not until Poe wrote *The Unparalleled Adventure of One Hans Pfaall*, in 1835, that the reasoned logical story of the interplanetary flight appeared.

Poe, however, founded this amusing adventure on the belief that the earth's atmosphere extends uninterruptedly to the moon. The method of conveyance was a balloon filled with a strange gas whose density is 341 times less than that of hydrogen — the lightest gas known.

Pfaall projects his balloon from the earth by setting off an explosive under it, and so provides an uprush of air that sweeps him aloft. After an amusing journey Pfaall lands safely on the moon and sends a message back to Rotterdam by a Lunarite, telling of his adventures.

The publication of *Hans Pfaall* at the same time that a similar story, *The Moon Hoax*, was appearing in a New York newspaper, served to raise a public controversy not only over the merits of the two stories but over the whole moon question. If a time were sought, then, for the beginning of modern serious interest in space flights and interplanetary communication the date of the publication of Hans Pfaall might be used.

Traces of the influence of Poe can certainly be found in the famous moon story of Jules Verne.[1] But Verne had not only half a century of additional scientific knowledge, but also a more scrupulous though less brilliant mind. Verne took the interplanetary journey, seriously, and tried so far as it lay in his power to make his account of it reasonable.

To project his explorers out of the influence of the earth and into the grasp of the moon Verne used a gigantic cannon with a bore 900 feet long, sunk in the earth, and pointing toward the zenith. Five hundred pounds of gun-cotton were exploded to shoot the space craft as a projectile into the void. One instant the craft was at rest, and the next instant it was plunging through the heavens with a speed of seven miles per second!

Many have questioned Verne's seriousness in subjecting the vessel and its occupants to such an instantaneous devastating impact. He must have known that the sudden acceleration of the ship would squash his passengers against the floor of the car as effectively as a fly is destroyed by a swatter. And if they survived, somehow from this disaster the friction of the meteor-like speed through the air would flash the ship, its passengers, and their dreams into flaming incandescence.

But his three explorers survive, and live to circle the moon and return safely to earth, assured, as a result of their journey, that the moon is indeed uninhabited.

Verne, like Cyrano, had the truth of interplanetary propulsion within his grasp

1 From the Earth to the Moon first published 1866.

without knowing it. He realized that his space ship, with no other power than the initial impulse, would no doubt crash on the moon after it crossed the border-line of attraction and escaped from the earth's to the moon's sphere of influence. He suggested therefore that to check the speed of the ship as it falls toward the moon rockets at the nose of the craft might be used!

It is a pity that he could not have seen that this very ability of rockets to retard the ship's flight could also start it, and that with the use of rockets there would be no necessity of starting the craft at a speed that would be absolutely destructive to the ship and its contents.

IV

Verne's story served as a powerful influence in the education of the public on the nature of interplanetary space and the scientific possibilities of space flight. His book had an enormous sale, and brought the question of interplanetary travel to the minds of millions of men and women. From the fanciful the space flight had become a problem of grim realities and terrible dangers; yet one that offered the possibility of the greatest adventure the race had conceived. It was Verne, truly, who prepared the public mind for the scientific proposals of Goddard, Oberth, and Esnault-Pelterie, that were to come five decades later.

Jules Verne expressed his feelings about the space flight through the words of his character Michael Ardan.

"In spite of the opinions of narrow-minded people," said Michael, "who would shut up the human race upon this globe, as within some magic circle which it must not overstep, we shall one day travel to the moon, the planets, and the sun with the same faculty and rapidity as we now make the journey from Liverpool to New York. Distance is but a relative expression, and must end by being reduced to zero."

These were stirring words for a mid-Victorian world; and there was one young man at least whose mind was inflamed by them and their suggestion of a new vision for a humanity weary of the wars, revolutions, and futile struggles of the nineteenth century.

In 1894 John Jacob Astor, writing in his *Journey in Other Worlds*, said: "We are all tired of being stuck on this cosmical speck with its monotonous ocean, leaden sky, and single moon that is half useless. Its possibilities are exhausted, and just as Greece became too small for the civilization of the Greeks, so it seems to me that the future glory of the human race lies in the exploration of at least the solar system!"

Astor disdained a trip as short as that to the moon. The giant Jupiter attracted him — its cloud-laden atmosphere and its great bulk suggestive of so much mystery and adventure. Placing his story in the year 2000, Astor felt himself free to utilize the inventions of that golden era, and his space ship therefore was propelled by a mysterious force known as apergy, which caused the earth to repel instead of attract the space-flyer — an angry mother thrusting away an ungrateful child.

In their twenty-five-foot cylinder Astor's explorers shot away from the earth, and after an uneventful journey in space landed upon Jupiter. They found the planet to be in the same stage of evolution as the earth during the Mesozoic period, harboring gigantic creatures, each a replica of some form of early earth life. From Jupiter the explorers traveled outward to Saturn, and, having traversed some two billion miles, they returned safely to earth.

Mr. H. G. Wells also used a gravity-repelling force for the space journey of *The First Men in the Moon*, written early in the present century. A substance known as Cavorite was conveniently discovered, which served as a gravity shield for the interplanetary sphere. Covered with Cavorite, the sphere had no weight, and could thus rise gracefully from the earth. Consequently, by unrolling the shield from the sphere on the side toward the moon, that body was allowed to exercise its attraction for the ship and to draw it gradually to her.

The discovery of a race of super-intelligent insects by the explorers provided Wells with an opportunity to paint his characteristic Utopia — a world freed from all confusing human emotions.

Faced with impending disaster on this new world, Bedford, Wells' hero, is led to examine the reasons for the exploration. "Why had we come to the moon?" he asks. "The thing presented itself to me as a perplexing problem. What is this spirit in man that urges him ever to depart from happiness and security, to toil, to place himself in danger, to risk even a reasonable certainty of death?

"It dawned upon me up there in the moon as a thing I ought always to have known, that man isn't made simply to go about being well fed and amused. Almost any man if you put the thing to him, not in words, but in the shape of opportunities, will show that he knows as much. Against his interests, against his happiness, he is constantly being driven to do unreasonable things . . . ."

CHAPTER III

THE MODERN IDEA - THE ROCKET

I

THE stories of space flights by Wells and his predecessors carry a practical admission that interplanetary travel is only a fanciful dream, no matter how desirable that dream might be. Wells and Astor used gravity-nullifying devices because there had been no other means invented that could conquer the earth's gravitation to ascend 240,000 miles of airless space to the moon, or 26,000,000 miles to Venus.

The aeroplane had just been successfully flown at the time Wells wrote his moon story, but both the aeroplane and the balloon were unfitted for the interplanetary journey. Yet even the first decade of the present century found men here and there convinced that a space flight could be made, although they had no vehicle.

The balloon must clearly fail, for it is an instrument of the air; its buoyancy gained only because it weighs less than the air it displaces. Even in the upper reaches of the atmosphere, where the air is exceedingly attenuated, the balloon is useless; and a practical limit for the balloon has been reached at about twenty-one miles above the earth's surface.

For the same reason the aeroplane must be discarded if we wish to reach interplanetary space. Not only is air necessary to the lift under the wings, to sustain it aloft, but it is necessary also to provide traction for the plane's propellers, and to supply its engines with oxygen. Although no definite height may yet be assigned for the limit of the aeroplane's field of action, it can hardly be used with efficiency above ten miles.

What was needed, then, was some machine which did not depend on air for its propulsion — one that could operate equally well in the vacuum of outer space as in our own atmosphere. Surely for the interplanetary enthusiasts of 1905 this was a large order. It meant something not only entirely new in operation, but also embodying revolutionary principles. It was as if a good merchant of the year 1750 had asked a carriage-maker to build him a conveyance to float through the air or under the sea!

It would seem, then, that those who believed, even against their own knowledge, that man would some day wing his way among the starry spaces were pursuing fools' fancies. Conservative opinion was emphatic on that point just as it advised would-be aviators of 1900 to desist from their efforts to build a heavier-than-air flying-machine. Besides, it was suggested why bother about interplanetary travel at all? What could it mean to us? The earth had troubles enough of its own.

To understand why small groups of isolated men kept alive the hope of reaching other planets (at a time when we could not ascend a hundred feet in an aeroplane) it is necessary again to turn back the book of history a few pages and to discuss the full significance of several important discoveries in nineteenth century astronomy.

One of the mainsprings of Verne's interest in the moon had been provided about thirty years earlier by a startling proposal of the director of the Vienna Astronomical Observatory. J. von Littrow was an eminent astronomer of his day, and his studies of the moon had convinced him that our satellite was inhabited. Why, then, he reasoned, could we not establish communication with the moon-dwellers by erecting an appropriate signal that they could understand and reply to by building one themselves?

Herr Littrow believed that a right-angled triangle would surely be recognized as a mathematical symbol by intelligent beings, and such a figure, seven to nine miles along a side, could be seen by Lunar astronomers if their telescopes were only as powerful as our own. He proposed, therefore, that this triangle be erected on the Siberian steppes and illuminated for the attention of the Selenites. Littrow offered as substantiation of his belief in a Lunar civilization the evidence of certain unexplained markings on the moon that one Gruithuysen, a "learned professor of Munich," called "a system of fortifications thrown up by Selenite engineers."

II

But Littrow, was unable to arouse enthusiasm over his project — the prevailing scientific opinion, since verified, maintaining that the moon could not support life. The world reluctantly turned away from the moon and focussed its desire for interplanetary communication on the 'war planet' Mars and on Venus.

Unlike the moon, these planets were believed capable of supporting life. And if that were true, reasoned imaginative theorists, why might not that life be similar to our own? Why should the earth alone be chosen to be the abode of intelligence? We had ceased to believe that the earth, or even the solar system, was made for and round us. Why, in short, might not we find on Mars and Venus life with which we could establish communication?

These thoughts, academic at first, became electric possibilities when Giovanni Schiaparelli, an Italian astronomer, discovered in 1877 what he called *canali* on the surface of Mars. What Schiaparelli understood as channels — straight lines some fifty miles wide and several hundred miles long criss-crossing the red planet — were accepted by the world as canals — the assumption being that they were artificial works.

At once intense interest was focussed on Mars; and Percival Lowell, the American astronomer, working at Flagstaff Arizona, devoted the remaining years of his life to an intensive study of these queer phenomena.

His conclusions will be mentioned in detail in another chapter, but it is sufficient to say now that Lowell died in the firm belief that the canals of Mars, so orderly in their arrangement, were the work of living beings — certainly of a high order of intelligence. To Lowell they were a system of irrigation canals a Herculean undertaking of the Martians to keep alive a dying world.

Lowell's conclusions were greeted by a wild flame of public interest, in which absurd statements were made both by his defenders and his opponents. He became a storm centre of controversy which raged into the present century, and which today is not entirely abated.

Now, and at last, reasons were found why an interplanetary voyage should be made. If Mars did hold living beings, perhaps like ourselves, why should we not communicate with them — for intellectual intercourse, for trade, for curiosity, and for the hundred other reasons that cause us to leave our peace and security for the danger of the new and unknown?

Scientists went further. If, as was believed, Mars is an old planet, one that is dying, then Venus, hidden by its perennial cloud layers, may be a young world and in a stage of evolution similar to the earth some twenty million years ago.

Whereas on Mars we ought to find a race millions of years in advance of ours, on Venus we might find life in the earlier stages of evolution. As Cavor said to Bedford in Wells' novel. "think of the new knowledge!"

And whatever doubts remained regarding the condition of our own moon, these too would be settled once a group of explorers landed there. In the minds of many scientists — astronomers, biologists, physicists. and chemists — there was reason enough now for an interplanetary journey. What was needed was a means of conveyance. As if by magic, the gift, in theory at least, was forthcoming.

III

Although the rocket was previously suggested by Cyrano de Bergerac and Jules Verne, both writers were ignorant of its potentialities. Probably the first man to realize its field of application for the space flight was Achille Eyraud, a writer of fiction and a contemporary of Verne. But Eyraud's book had none of the sensationalism of Verne's, and because it dealt with an instrument that was not spectacular it attracted little popular interest. But two serious men of science at the dawn of the present century were running over in their minds the possibilities of the rocket.

Robert Esnault-Pelterie a French engineer, was intensely interested in the new aeroplane and the motor-car just having their debut. He has, in the words of one writer, "one of the most original technical minds in France," and when not inventing new and revolutionary equipment for aeroplanes and motor-cars he was engaged in examining the physical problem of finding a means for transportation at high altitude and lifting bodies from the earth to outer space by the rocket principle. His first calculations were completed in 1907; he amplified them, and presented his conclusions to the Societé Française de Physique in 1912.

That scientific interest in an ascension from the earth higher than man had ever gone was awakening can be judged also by the granting of a Belgian patent to a Dr. André Bing in the previous year, for "an apparatus to permit the exploration of the high regions of the atmosphere regardless of rarefication of the atmosphere."

The year 1912 found another original scientific mind, that of Dr. Robert H. Goddard of Princeton University, making calculations on the possibilities of the rocket. Goddard's findings were so interesting that when he later went to Clark University at Worcester, Massachusetts, he continued in 1915 and 1916 to test them by a series of careful experiments.

He published his conclusions in 1919, in a comprehensive report to the Smithsonian Institution,[1] and stated that it was his earnest belief that the rocket could be used as an agent to penetrate the high rarefied layers of our atmosphere, where the balloon and aeroplane could not be used, and in the vacuum of interplanetary space. By a series of ingenious inventions of various rocket devices Dr. Goddard laid the foundation for the development of his ideas into a practical vehicle, revolutionary in its implications.

1 Robert H. Goddard, A Method of reaching Extreme Altitude (Smithsonian Miscellaneous Collections, vol. lxxi No. 2).

Almost immediately afterward came the publication of researches of a noted Austrian, Dr. Hermann Oberth, and the German Dr. Walter Hohmann. What was significant in the results of the work of these men was their unanimous conviction that the rocket could not only be used in our upper atmosphere, but that it could also make a journey to the moon, to Mars, Venus, or wherever we had fuel and supplies to take us.

In substantiation of his beliefs Oberth completed in 1923 an extensive mathematical and technical survey of the problem, to be extended still further in 1928. In France Esnault-Pelterie, stimulated by his previous calculations, completed in 1927 his own work, *L'Astronautique*, a gigantic monument of scientific research and planning.

Such was the startling series of developments that awakened the twentieth century to the realization that a new era of exploration was dawning. Thousand mile-an-hour rocket-planes and interplanetary travel were no longer the butt of the satirist or the tool of the novelist. They were worthy of the study of men of science, who were putting behind their conclusions the weight of their reputation and authority. At once, chiefly in Germany, a thirst for knowledge on the subject arose — and to meet it men like Max Valier and Otto Willi Gail responded.

Valier, a keen student of science, threw himself with supreme devotion into the task of convincing Germany that the rocket-plane and interplanetary flight were possible. Writing, lecturing, and experimenting continuously, Valier became known to every German.

No less enthusiastic was Otto Willi Gail. A close friend of Oberth, he saw the scientist's vision, and in the two novels *The Shot into Infinity* and *The Stone from the Moon*, Gail gave us a picture of an Oberth combating prejudice, opposition, and treachery to build the first space-flyer for man's greater glory.

Around such guiding lights as Oberth, Gail, Valier, and others the German Interplanetary Society was organized. Among its thousand members it numbers many of the keenest scientific minds of Germany.

In France a similar organization for the study of the interplanetary question was arising. Dominated by an unselfish desire to promote what he believed to be a great adventure and conquest for man, Esnault-Pelterie, with his friend André Hirsch, established the Rep-Hirsch Fund, which now awards 10,000 francs each year to the author of the most original contribution to the new science of navigating the interplanetary spaces — astronautics. And to aid them in their selection the Committee for Astronautics, a group of eminent French scientists, was formed.

In America popular interest in the problem was slow in awakening, and no definite organization was created until the spring of 1930, when a group of scientists and laymen founded the American Interplanetary Society. In Russia too, under the leadership of Professor Nikolas Rynin, a literature on rockets and their uses assumed scientific importance.

Experimentation on the rocket, to find the means to harness most effectively its great gifts, is proceeding with vigour today in both America and in Germany, under the leadership of Dr. Goddard in the United States and of the German Interplanetary Society in Germany. And at the present time plans for an International Interplanetary Commission, at the suggestion of the author, are being devised to unite the rocket-experimenters of the world into a federated group. These movements are all to be expected to hasten the day when the rocket will enable man to span the higher altitude and bring the distant planets within his reach.

CHAPTER IV

THE ROCKET AND ITS DEVELOPMENT

I

The simplest analogy to the rocket is found in the action of firearms, or in the skyrocket used in celebrations and warfare.

Everyone who has handled a rifle is familiar with the recoil that accompanies the firing of the gun. Newton explained this recoil on the principle that "action is equal to reaction," which, in lay terms, means that the force in the powder that propels a bullet on its way is equal in every respect to the force that kicks the gun back against the marksman's shoulder. And if the rifle is not held motionless by the muscular force of the marksman it will actually travel backward as the bullet goes forward. This action is apparent to anyone who has witnessed the firing of large ordnance — the gun rolling backward on its wheels with considerable violence as the shot is fired.

To understand this action fully we must examine the nature of the phenomena accompanying the firing of a gun. The powder used for propelling the bullet is exploded by the action of the trigger on the cap. The detonation of the powder flashes it into a gas whose volume is hundreds of times greater than that of the powder — thereby creating a terrific pressure in the shell.

This pressure acts equally in every direction. It not only acts on the side walls of the shell, but also along its length. The force parallel to the shell pushes the bullet out of the gun, giving it a great velocity, and it presses back against the cap end of the shell. This latter pressure is transmitted through the gun to the shoulder of the marksman. Although the force that sends the bullet on its way is the same as that which causes the recoil, the bullet is given greater velocity because its mass is so much smaller than that of the gun.[1]

This principle was applied by Goddard and others to the rocket. They conceived a metal projectile-like instrument containing a chamber, one end being open to the outside (air or vacuum), in which a fuel would be exploded. The resulting gases created by the explosion expanded with great energy, pressing against the front wall of the chamber, and were expelled into the open. But in this process of expansion and pressure against the front wall they pushed the rocket forward in a direction opposite to that of the expulsion.

Here was an action of the most compelling simplicity, yet one that on close study was capable of developing enormous power. Free from moving machinery or complications of equipment, the rocket, in theory at least, could make full use of the enormous power latent in explosive fuels such as gunpowder. And if this rocket motor, as it is called, were attached to a vehicle of any kind that vehicle could, again in theory, be propelled forward at almost cosmic velocity.

II

When the first popular works on the rocket appeared skeptics on its power arose on every side. "How," they all exclaimed, "can this rocket, which depends for its action on the expulsion of gases, operate in the vacuum of interplanetary space, where there is no air for it to press upon? How, in short, is this means of propulsion to fulfill the dream of flight to other worlds?"

Those who do not and will not understand the action of the rocket abound everywhere. The propulsion of the rocket, as was stated, takes place in the combustion chamber because of transmission of the energy of the expanding gases to the front wall of the chamber. Since all the work is done by the gases before they are expelled this action is independent of the medium in which the rocket moves. And as has been proved by experimenters, the rocket is actually more efficient in a vacuum than in a space filled with air. A little explanation is necessary to clear up this doubtless paradoxical statement.

1 "Force" = "mass" x "acceleration." Since the mass of the bullet is considerably less than that of the gun it is obvious that to have equal and opposite forces at the instant of firing the velocity of the bullet, which is a function of the acceleration, must far exceed that of the recoiling gun.

The forward motion of the rocket, as we have seen, is necessitated by the pressure of the expanding gases, and this process, as defined by Newton, is one of action and reaction — the gases rush in one direction, the rocket in the other. Any hindrance to the expansion of the gases and to their subsequent expulsion from the rocket chamber would thus reduce the reactive effect and reduce the energy transmitted to the rocket. If the exit of the combustion chamber were thus sealed up, so that no gases could escape, it is evident that there would be no movement of the rocket.

To put it plainly, the mass of the gases multiplied by their speed of ejection will equal the mass times the speed of the rocket. The speed of the rocket will therefore depend upon the speed of the expelled gases.

The rocket, moving through the atmosphere, expels its gases into the air, whose back pressure hinders their exit from the chamber, thus reducing the velocity of ejection. In a vacuum there is no such hindrance, and the ejected gases rush into space with a maximum velocity, and therefore with a maximum transmission of energy to the rocket.

A second factor also proves the rocket to be most efficient in a vacuum. The rocket is an instrument for the attaining of exceedingly high speeds for, as recent studies have shown, at low speeds it is inefficient. At such speeds as a mile a second and upward the resistance of air to the passage of the vehicle through it would be terrific, and the power necessary to combat this resistance alone would be colossal. In a vacuum, on the other hand, there is no such resistance; there is absolutely nothing except gravitation to retard the flight and the attaining of speeds that on earth would seem impossible. As a result, the exponents of the space flight are unanimous in stating that in a vacuum the rocket develops its greatest power.

With the use of the proper fuels, therefore, and the absence of the resistance that limits speed in our atmosphere, the rocket in airless space can with ease attain the speed of six to seven miles a second necessary to escape the earth's gravitation and to fly to the moon, Mars, or Venus. Or it can, as a necessary prelude to this ambitious project, be used as a means of transportation between distant terrestrial centers of population. Ascending into the upper reaches of the atmosphere, where the air is exceedingly rare, rocket-planes can attain, in this area of little friction, speeds of three thousand miles an hour and upward, and bring Europe and America within one hour's journey of each other.

III

Thus far we have been dealing with the rocket as if it were a new invention, originated in the minds of men purposely to enable them to achieve the conquest of space.

The rocket on the contrary, is a heritage from past ages, and comes to us still in a crude stage of development. Although it has been known to the civilized world for a thousand years, the astronauts of the present century, when they seized upon it, realized that they had only a principle of motive power, and not an engine ready for the tremendous task of the space flight.

As a "firework;" the rocket has been used for hundreds of years. A simple paper tube constricted near one end and filled with gunpowder is ignited, and the expanding gases send the rocket aloft. Showers of beautiful sparks or elaborate designs in light are released from the head of this skyrocket while in flight, and at public celebrations the rocket brightens the night sky with dazzling displays of blended colors.

The invention of the rocket may no doubt be attributed to the Chinese, the inventors of gunpowder. As an engine of war the Chinese are said to have used the rocket against the Tatars in the thirteenth century. By filling the head with inflammable material they shot against the enemy a terrible projectile that spread masses of hungry flames.

It was not long before its ability to rise high into the air and to discharge a cascade of sparks, or a charge of magnesium emitting an intensely bright light, led to the rocket's use in Europe as a means of signalling and of discovering movements of an enemy in warfare. But its practical value was quite limited. The fuels, encased in inflammable paper, were apt to burn up the rocket at the moment of ignition; and if they were started successfully there was no way to guide them except by means of a long stick attached to the end. If the wind were unfavorable, or the stick badly placed, the rocket might turn suddenly like a boomerang, to spread devastation in the ranks of its senders.

It was not, in fact, until early in the nineteenth century that the rocket achieved a definite development unfortunately as an engine of warfare. In the hands of Sir William Congreve, an English ordnance expert, it was used with amazing success, there being occasions when it almost appeared that it would supersede as a means of deciding the Napoleonic wars.

Congreve took the rocket, unchanged for centuries, and gave it a case of iron instead of paper. He perfected its internal construction — the size and shape of the combustion chamber and the exhaust nozzles — and built rockets larger by far than had ever been used before. He filled his rocket-heads with lumps of iron and inflammable materials, and shot them against the French town of Boulogne in 1806. The rockets burst upon landing, generating heat intense enough to burn up the covering shell, and exposing a great mass of flaming material to burn up anything inflammable which it touched.

The Boulogne bombardment by rockets was extremely successful. Since rockets needed no external means of propulsion, such as heavy ordnance, Congreve was able to fire a number of them with great rapidity from rowboats in the waters off Boulogne.

Despite Congreve's successes, however, the rockets had definite disadvantages in warfare. They were difficult to aim, and with only a stick to control the flight there was no predicting where they would land. Changes of temperature often caused the cases to expand or to contract, thus shifting the pressure of the charge, with the result that they often burst at the moment of firing. Then, too, they were dangerous to handle at the moment of filling with fuel, and in storing — and soldiers and sailors soon acquired an unconquerable distrust of them.

The middle of the nineteenth century saw the rocket practically obsolete as an arm of war. Artillery itself had been enormously improved in the interim, and the relative advantages of the rocket were no longer the same in 1850 as in 1807. With the precision in firing attained by rifled artillery in the middle of the century the rocket was definitely supplanted, and by 1870 it was a thing of the past.

Improvements, however, were not lacking. The great problem of stability in flight was achieved in part by the use of inclined shields, or flanges, at the base, instead of a mere stick. The escaping gases pressed on the flanges, causing them to rotate or spin, and giving the effect, in part, of a rifled shell.

With increasingly better construction and better materials, the rocket, shunned by armies, came to be accepted as the standard equipment for coastal lifesaving stations. Carrying ropes, they often solved the difficult and important problem of establishing physical communication between wrecked ships and the shore, and were instrumental in saving thousands of lives.

Their use was further facilitated by the invention of the double rocket by an army officer — a Colonel Boxer. By placing two rockets end to end in a case Boxer found that the first would, on ignition, carry the projectile upward to the limit of its charge, at which time the second would become ignited and provide a further impulse. This double rocket, or "step-rocket" as it is called, was destined, as we shall see, to become one of the most important developments in the programme for the conquest of interplanetary space.

With the advent of the World War the rocket again became an important military arm, but chiefly as a means of releasing a magnesium flare for signalling and for illuminating "No Man's Land." Its use as a life-saving device had been extended constantly, until in 1922 in Great Britain alone 300 coastal lifesaving stations were equipped with rocket batteries.

A device, however, to carry a shell two or three miles, or a life-line half that distance, is a far cry from the construction of a rocket that can rise 50 to 250,000 miles above the earth.

Although the mathematical calculations of Goddard, Esnault-Pelterie, and Oberth showed that the rocket could fly into interplanetary space, these scientists have agreed that years of experimentation must first ensue. The chief problem is that of an adequate fuel — one which for each pound of its own weight will release on ignition a definite amount of available energy. The necessity for a determined minimum of energy from the fuel was fundamental in the calculations of these men.[1]

The rocket, unlike the aeroplane or rifle shell, must carry all its fuel for propulsion. The fuel must lift not only the weight of the ship against the gravitation of the earth, but must also lift its own weight. Thus if an unduly large amount of fuel were required to shoot the ship into interplanetary space, more fuel must be carried to lift the fuel, ad absurdum.

The fuels that had been used in warfare and for life-saving lines were totally unfitted for the giant work of propelling a space-flyer, and it was recognized that a new science of rocket fuels must be developed.

Since little was actually known of what power fuels could develop in the rocket Dr. Goddard performed a primary series of fundamental experiments at Clark University in 1919 under a grant of funds by the Smithsonian Institution, the results of which were highly encouraging. At that time Dr. Goddard did not express the desire to fly into interplanetary space — for even in 1919 that was too much of a fool's dream for the conservative Smithsonian to sponsor. His aim,

1 It was calculated for example, that the lifting of one pound of weights to the limit of the earth's gravity (with relation to the moon), or to send one pound, away from the Earth at seven miles per second, required the entire energy of seven pounds of pure carbon and oxygen.

he stated, was to develop a rocket that could penetrate our atmosphere to heights never before attained, and by the instruments for measurement and calibration that would be carried gather valuable information about the nature of the upper air.

The 1919 experiments of Goddard were static or laboratory affairs — to measure with exactitude the power developed in the rocket by various types of gunpowder. It was then that Goddard verified definitely his belief that the rocket would operate as effectively, or more so, in a vacuum as in air.

Although Goddard's conclusions, published in his report to the Smithsonian Institution, revealed what might have been startling facts to the world, public opinion would not yet recognize their implications. Confined to a slender grant of funds, Goddard could do nothing but work patiently in his Worcester laboratory at his designs of rocket shapes and fuel compositions, inventing newer and newer devices to utilize the rocket principle, and attempt meanwhile to educate the public to the significance of rocketry.

He affirmed his belief that it would be possible with further development of the rocket to send a shot to the moon, and by the explosion of a charge of magnesium powder on landing a brilliant flare would be set off that would indicate to watchful astronomers at their telescopes that man's messenger had conquered the great spaces. The successful termination of such a flight would be, of course, the first proof that a space-ship could be built to carry living things.

The publication of these statements served at once to make Goddard a target for most violent attacks.

Even the ability of the rocket to operate in airless interplanetary space was seriously questioned by men of assumed scientific training; while the attaining of the tremendous speeds necessary to escape from the earth's gravitation was considered the height of absurdity. A Mr. Morell, writing in the London Graphic of November 20, 1920, stated that the "rocket will generate a red heat for the first hundred miles of its flight through the atmosphere"; and travelling at six to seven miles per second, as it must to escape from the earth, the rocket will "vanish in an incandescent wisp of flame and smoke."

To these and similar criticisms Goddard replied patiently, avoiding, as he said, the "sensationalism" that was inherent in the thought of a moon flight.

He pointed out gently that to avoid the devastating friction of the air rockets would be sent through the atmosphere at a safe but gradually accelerated velocity, and would not achieve their seven miles per second until the vacuum of interplanetary space had been reached, and no friction was possible.

With the opening of the third decade of this century, therefore, the idea that the rocket was a practicable principle for terrestrial and interplanetary transportation was being impressed forcibly on scientific minds.

CHAPTER V

TAMING THE GIANT

I

THE first series of experiments made by the German group of rocket enthusiasts were attempts to "try out" the peculiar characteristics of the rocket as a means of propulsion. It was natural, therefore, that it should be applied to terrestrial vehicles such as the motor-car. Succumbing to the eloquence and persuasiveness of Max Valier, Fritz von Opel, a wealthy young sportsman and motorcar builder, became intensely interested in the rocket, and set out to probe its possibilities.

On the Avus Speedway in Berlin in the summer of 1928 he used rockets on the tail of an automobile, and achieved a sixty-mile-an-hour speed within 200 feet of his starting-point! Such extremely rapid acceleration had never been heard of before. This, together with the fact that he finally achieved a speed of 130 miles per hour before his rockets were used up, convinced Opel and others that here was an instrument capable of delivering enormous quantities of power in a relatively short time.

Opel's rockets consisted of powdered fuel enclosed in separate containers, each container burning continuously until the fuel was exhausted.

Encouraged by his results, Opel successively built a second, third, and fourth car. And when the fourth was blown to bits early in August 1928 this dauntless man determined to start work at once on a fifth. But under the stricture of official disapproval he was compelled to desist.

Max Valier, meanwhile, was making his own experiments. On July 18, 1928, he made three tests of rockets, placing them on a light wooden car. With the first two he achieved a speed of 112 miles per hour, while in the third, with a speed of 131 miles per hour recorded, the light car, unable to stand the terrific pace, overturned and was wrecked.

Unafraid of this giant of power that he was attempting to tame, Valier applied the rocket next to drive a sledge over the ice of Lake Starnberg, in Germany, and attained a velocity of 235 miles an hour.

If anything were needed to cap the climax of Valier's enthusiasm for the rocket this test supplied it, and he confidently predicted that 300 miles an hour was the next step. He believed, further, that he could build an aeroplane that would operate by rocket propulsion; and planned a flight from Calais to Dover that very summer.

Although the experiments already made indicated conclusively that the rocket was, in fact, an engine of unusual power, dissatisfaction with its operation was quite universal among authorities. The disastrous results of several of the experiments indicated that the powdered fuel was difficult to control. The energies available, although great, were deemed insufficient to propel a ship into interplanetary space. Calculations of Goddard and others had shown that to lift a one-pound weight beyond the earth's gravitational influence would require at least 400 pounds of the best powdered fuel. The solution of the space flight upon this basis was palpably impossible.

Attention shifted abruptly, therefore, from powdered fuel, and research was begun for something that would not only offer a far greater amount of energy per pound of its weight, but that could also be controlled with greater safety. As it possessed both of these qualities, liquid fuel was suggested in several quarters, and it was seized upon as the ultimate means of propelling terrestrial and interplanetary rocket craft. A mixture of liquid hydrogen and oxygen, for example, called "detonating gas," could provide a continuously burning propellant (whereas powdered fuel had to be used in separate containers, with a corresponding increase in weight of containers), and has an inherent energy some three times greater than the best smokeless powder found by Goddard.[1]

The Breslau Society for Aerial Navigation officially sanctioned the use of liquid oxygen and alcohol, and the fuel was accepted by Hermann Oberth, then engaged in the construction of a giant test space rocket, as the basis of his designs. In America Dr. Goddard had independently turned to it; and, writing in his L'Astronautique, Esnault-Pelterie had used it as the starting-point for a discussion of the mathematics of the space flight.

The question of fuel is dwelt upon thus fully because it is still the basis of all

1 The inherent energy of combustion of a pound of "detonating gas" is approximately 6900 B.T.U., as compared with 4000 for pure carbon and oxygen. (27,000 B.T.U. is necessary to lift one pound beyond the earth's attraction.) The figure of 6900 of course is the energy available with 100 per cent efficiency, which is not realizable in practice. Of almost equal power is a mixture of liquid oxygen and alcohol.

discussion on rockets. The power necessary to lift a body away from the earth and give it a speed of seven miles a second, in order to free it from the earth's attraction, is truly enormous, and none recognize that more than the space flight's exponents.

II

While believers in rockets abroad debated these questions in public Goddard, in the quiet of his Worcester laboratory, went unobtrusively on with his work, preparing his most ambitious experiment — this time with a liquid fuel.

With two decades of rocket research behind him, the Worcester physicist realized that the elements of the rocket must be placed under control before it could become a practical instrument.

Fuels must be found that provided, per pound of their weight, energy enough to lift a vehicle against the earth's attraction to a desired height. Secondly, means must be devised to control the action of the fuels — the volatile liquefied gases, hydrogen and oxygen, must be safely stored in the rocket and fed into the combustion chamber so that they would not only be under control at all times, but also give up the maximum of their energy to the rocket's propulsion. Thirdly, the rocket must be designed so that it would be stable in its flight.

To test a design calculated to meet these difficulties, and also to determine if a rocket could ascend to a respectable height carrying delicate instruments with safety, Goddard planned his 1929 experiment.

He prepared a parachute in his rocket, its release from the nose at the upper limit of the projectile's flight permitting the rocket to drift back to earth without damage to the instruments. If he could succeed Goddard had in his hands a powerful argument for the obtaining of greater financial support for his experiments than was possible under his slender Smithsonian grant. Such a rocket sent 10, 20, 50, or 100 miles into the air, and recording temperature, air-pressure, wind movements, and solar activity, would be of the greatest value not only in weather-forecasting, but to aviators the world over.

For his purpose Goddard built a rocket nine feet long and twenty-eight inches in diameter, made of metal, to be propelled by a continuously burning fuel, composed of liquid oxygen and an unnamed hydrocarbon. A steel tower forty feet high was erected, with metal grooves along which the rocket was to be guided on its upward flight.

The firing of the rocket into the sky outside Worcester startled the residents of the town, but not more than it startled the newspaper reporters present. Excited reportorial imaginations pictured Goddard attempting secretly to shoot a rocket to the moon. Since there was no evidence of the success of a moon flight the rocket had "exploded," thereby demonstrating to the world that the rocket flight was like Wright's aeroplane in 1903 "in the class of perpetual motion."

But it was not until the first excitement of this shot had worn off that the news finally filtered to the public that no moon flight had been contemplated, and that nothing had exploded. A shot of several hundred feet into the air had occurred and the instruments attached to the rocket had come through the flight uninjured, making the experiment a complete success.

It was probably this flight of July 17 that operated to attract to Goddard's work the interest of the late Daniel Guggenheim.

Not quite one year after the July 17 experiment Guggenheim announced the donation of a sum equivalent to $100,000 (£20,000) to the Goddard experiments (after what must have been the most careful investigation) — and with it the support of a board of America's most eminent scientists. This grant, more ambitious in its implications than the Smithsonian fund, had for its purpose the construction of a rocket to penetrate beyond the atmosphere, some 250 miles above the earth, and to record with instruments information about the upper air and the vacuum of space.

The thirty years of effort by Goddard, Oberth, Esnault-Pelterie, and others were bearing fruit. America was shaking off its skepticism and beginning to take the rocket seriously!

III

The excitement aroused by the Goddard experiment of July 17 had hardly died down when the world was startled by the announcement of a successful flight of an aeroplane propelled only by rockets. On September 30, 1929, using a powder propellant, Fritz von Opel was shot along a track in Frankfurt, and, rising about forty-nine feet into the air, flew for one and a quarter miles before the plane landed as a wreck.

He achieved a speed of 85 miles per hour, controlling the flight by the timing of the rocket shots. When one considers the lack of control over powdered fuel and the fact that the track was short, and did not allow for maneuvering, the audacity of Opel's flight can be appreciated. As the first aeroplane to fly under power

CHARGING A ROCKET WITH FUEL AT THE
"RAKETENFLUGPLATZ"

other than that of a petrol engine, the Opel plane opened up to aviation a new perspective, revolutionary in its implications. It further added weight to the force of the Goddard experiment in focussing public attention on the unusual qualities of the rocket.

With the close of 1929 the "rocket proposition," as it was called, was thus in the most hopeful state since the idea had been first proposed. Although on the surface the work of Opel and Goddard had failed to disturb the mainstream of twentieth-century life, there was an undercurrent of hope and speculation. There were new aspirations arising everywhere.

It seemed too unreal to be true — the very vividness of the stories that had been written about interplanetary flights seemed to throw over them a glamour of the fancy that could not be dispelled. The primitive man's fear of the heavens and the terrible emptiness of space reasserted itself in a world of rationalists; people feared and expected anything.

Something did happen, five months later, to fulfill these fears. It seemed that the curse of Phaethon was being renewed in the enlightened atmosphere of the twentieth century.

In the spring of 1930 Max Valier had driven a motor-car propelled by a new rocket motor, and with it he had developed a speed of 60 miles an hour. The motor, using a fuel of liquid oxygen, water, and methylated spirits, and designed by Dr. Paul Heylandt, weighed only seven pounds, and achieved surprising power. The 40 or 50 horse-power developed was epoch-making in its significance, for it was five times as much as that of a motor-car engine of the same weight, and six to seven times as powerful as the best aeroplane engine.

At last an approach was being made to the astronaut's requirements for fuel energies. A seven-pound motor providing 50 horse-power was not yet powerful enough to shoot a ship beyond the earth's attraction, but it was a sign that the problem was being progressively solved.

A month later, supervising the installation of an improved design of the Heylandt rocket motor on a car, Valier, while standing behind the car to test the pressure and capacity of his fuel container, was blown twenty feet into the air by a sudden tremendous explosion, followed by a flame ten feet long that projected instantly from the tank.

Thus astronautics, the new science of man's conquest of space, claimed its first victim, and a choice one.

This man, who at the age of thirty-five had done more than any other to make a nation "space conscious," was paying the penalty for tampering with the gigantic fires locked in Nature's cupboards.

Yet it is significant that at the very hour that Valier's life was being mercilessly snuffed out Esnault-Pelterie was convincing a group of French scientists assembled at Paris that a moon flight could surely be made within fifteen years!

IV

Experimentation in the development of the rocket since the tests of 1928 and 1929 have indicated a growing control and increasing skill on the part of astronauts. Whereas the earlier work was designed almost exclusively to test the rocket's peculiarities in operation, the more recent efforts have been devoted to building a projectile that will actually fly. It was evident that for this end a long series of scientifically planned experiments would be necessary.

The German Interplanetary Society had already made tentative experiments on liquid fuels, and was now ready to continue them in a more ambitious manner. In the autumn of 1930 it obtained a tract of land outside Berlin, and there established the first Raketenflugplatz or "rocket-flying field."

I quote from Mr. C. Edward Pendray, Vice president of the American Interplanetary Society, who visited the field in the spring of 1931. It is, he says, by far the most extensive experimental ground for the study of rockets in the world. It lies in Reinickendorf, not five miles from the heart of the German capital, and sprawls northward into the hilly, tree-protected country outside the metropolis. It is larger than the famous flying field of Tempelhof, and on it a full-time staff of six engineers are working seven days a week to accomplish the miracle of rocket transportation . . . .

Before this comes to pass, however, a stupendous amount of work must be done. We are still in the first stages of rocketry, that of sending up simple rockets to high altitude, and a large part of the work is now being done not with actual rockets, but on what technicians call the proving-stand a 'set up' on which rocket motors can be tested as to lift and efficiency without going to the expense and trouble of building the entire rocket. Less spectacular than actual rocket shots the proving-stand work is nevertheless extremely important at this stage.

At the rocket-flying field an elaborate technique has been worked out for testing rocket motors on the proving stand. This work is necessarily dangerous, and every precaution is taken to have all workers in safety behind embankments during rests. The fuels are turned on by remote control, and the lift of the motor is automatically recorded by a special clockwork device.

Like the American Dr. Goddard, the engineers of the Raketenflugplatz look not to sensational immediate results, but toward the building of a technique of rocketry that will make its future secure.

Mr. Willy Ley, of the German Interplanetary Society, writing to the author, has this to say:

> Important rocket experiments have been performed by the technicians of the Society, the engineers Nebel and Riedel — experiments dealing with the combustion of petroleum with liquid oxygen, the stabilizing of the rocket in flight, the arranging of the fuel chambers, and of the parachute which will bring the burned-out rocket safely to earth.
>
> It is a great pleasure to say that all these experiments, despite their great difficulty, have been successful.

As a result the German Society is building at the flying-field a rocket expected to ascend some three miles into the air.

V

The early spring of 1931 saw a renaissance of interest by German experimenters other than those laboring at the rocket-flying field, and the recording of new heights for rocket ascension.

Shooting a test rocket at Dessau on March 14, 1931, Johannes Winkler, the founder of the German society, watched it ascend a thousand feet into the air, to land gently upon the earth not six hundred feet from the starting-point. Propelled by liquid oxygen and petrol, the rocket was but two feet long and one foot in diameter, and was ignited electrically from a distance of 150 feet.

The rocket of Karl Poggensee, a German aviation engineer, rose even higher in a test near Berlin on March 13, 1931; it ascended 1500 feet, carrying its delicate measuring instruments, consisting of an altimeter, a camera, and speed-measuring devices, and landed them by parachute without damage.

On April 15, 1931, a height of 6000 feet was reached by a series of rockets loosed at Osnabruck by Reinhold Tiling. Six rockets were shot upward, of which one exploded at a height of 500 feet. The others, attaining a speed of 700 miles an hour, rose over a mile before their upward impulses were lost. Herr Tiling used, instead of the familiar parachute to bring the rockets safely to earth, a fundamentally new device — small blades which were automatically unfolded by a clockwork when the upward momentum had disappeared, thus making of a descending rocket a virtual autogyro.

More ambitious in size also than previous rockets were the Tiling projectiles, one being fifty-four inches long, with a wing spread of eighty inches. Tiling contemplates the building of a rocket to ascend 60,000 feet, with a wing spread of forty feet.

CHAPTER VI

ROCKET PROBLEMS

I

The problem of constructing a rocket to ascend high into the air, and whose flight is to be controlled, demands for its solution great skill and ingenuity. Whether or not it is to carry passengers, whether its range is to be 50 or 50,000 miles above the earth, it must be considered both a vehicle and a projectile. And unfortunately this two-fold division often calls for diametrically opposed designs.

Since the rocket must combat the gravitation of the earth the construction must be as light as possible. Because its parts will be subjected to great stresses in flight it must be designed and built for considerable strength. As a projectile it must obey the laws of ballistics, and the weight and location of its component parts must be balanced so that it will not be top-heavy or awkward in flight. Travelling at relatively high speeds, it must be stable in flight and admit of directional control. Finally, since the power required to send it high above the earth will be great, it must be an efficient engine, changing fuel energy into propulsive force with a minimum of loss.

The rocket consists fundamentally of four parts

(1) the fuel compartments and their auxiliary equipment; (2) the combustion chamber, in which fuel is burned; (3) the pay load compartment, containing the equipment, passengers (if any), recording apparatus, and landing devices such as the Parachute; and (4) the rocket shell.

The general rules that must be observed in the construction of each of these parts and their interrelation are practically independent of the purpose of the rocket.

The particular design of a rocket, however, will depend fundamentally upon its purpose and upon the fuels to be used.

Two types of fuel have been experimented with to date — solid fuel, such as gunpowder; and liquid fuel, such as the combination of liquid hydrogen and liquid oxygen.

Although both types are still used in rocket research, experimenters have been convinced that the liquid fuel not only offers considerably more energy per pound, but admits of greater control and a wider range of usefulness. The future of rocket propulsion, they are agreed, especially for vehicles that are to rise to a considerable height, rests upon intensive experimentation with the most powerful types of liquid fuels.[1]

II

The combustion chamber is the heart of the rocket, and its requirements are perhaps the most exacting of all. This chamber — referred to as the motor — receives the fuel. In it the fuel is burned, the ejection of the resulting gases creating the reactive force that propels the rocket. The chamber must be constructed, therefore, so as to permit the proper introduction of the fuels, to provide that they meet in a manner to make combustion as complete as possible; and to force an ejection of gases with the greatest possible speed.

In the skyrocket, a dry or solid fuel rocket, these problems are comparatively simple. The fuel, gunpowder or a similar explosive, is a permanent coating on the combustion chamber walls, There is naturally no problem of introducing it. Since the powder needs no second material for combustion there is no problem of mixing fuels.

1 There are two criteria by which fuel's value to a rocket may be tested. The energy released in combustion is a measure of the available energy for propulsion; and the ejection velocity is an index of the propulsive force that may be obtained.

Since the reason for the second standard may not at first be clear a few words of explanation may be given here. The reactive force in the burning of fuel is obtained by the ejection of the gases from the rocket. As the gases are thrown backward the reaction moves the rocket forward. From physical principles the total momentum imparted to the rocket is equal to the momentum of the ejected gases; and since momentum is a product of mass times velocity it follows that the mass of ejected gas multiplied by the ejection velocity will equal the mass of the rocket times the velocity imparted to it.

Thus it follows that one pound of fuel ejected with a velocity of 1000 feet per second will give to a ten-pound rocket a velocity of 100 feet per second. If the ejection velocity of the fuel could be increased to 2000 feet per second, it should be clear that the velocity of the rocket would be 2000 feet per second.

Liquid fuels not only have an available energy of several times that of dry fuels but the ejection velocities obtained by liquid fuels are considerably greater.

The combustion chamber of the skyrocket is simply constricted at the exhaust end, so as to create great pressure of the resulting gases in the chamber. A wick is inserted into the fuel and lighted. The fuel gradually burns away, turning into gas, which on ejection sends the rocket into the air. As the fuel is consumed the chamber becomes larger. When the fuel is exhausted the impulse ceases, and the rocket falls back to earth.

Attempts have been made to extend the use of the dry-fuel rocket beyond the skyrocket. Much of Dr. Goddard's early work was with dry fuel, and Fritz von Opel uses a powdered fuel to propel his first rocket-plane. The experimental rockets shot up 6000 feet by Reinhold Tiling also used dry fuel. But in each case the method of combustion was similar to that of the skyrocket.

Opel and Tiling used a number of separate unit rockets, while Goddard planned for a number of fuel cartridges to be sent into the combustion chamber.

In other words, in no case was there a continuous flow of the fuel.

Attempts to equip a rocket with a central reservoir of dry fuel, and to pump it into the combustion chamber in a steady stream, have failed. The powder could not be adequately controlled, and it was found that the pressure or friction upon the fuel, while forcing it into the combustion chamber, often caused premature explosions. The dry-powder rocket is therefore limited to designs in which the powder is already in the combustion chamber, or fed into it by means of cartridges. Its use is naturally limited to conditions where the maximum of energy per pound or maximum ejection velocity is not necessary.

III

The liquid-fuel rocket offers infinitely greater possibilities, but at the same time presents many new problems. Since the burning fuel in the combustion chamber exerts great pressures throughout the chamber additional fuel, to maintain continuous burning — i.e. continuous reactive force — must be pumped in under pressure. The shape of the chamber must also be designed to allow for maximum pressure and the most complete combustion of the fuel.

The German engineers at the Raketenflugplatz have been working energetically on the construction of a combustion chamber of the greatest efficiency.

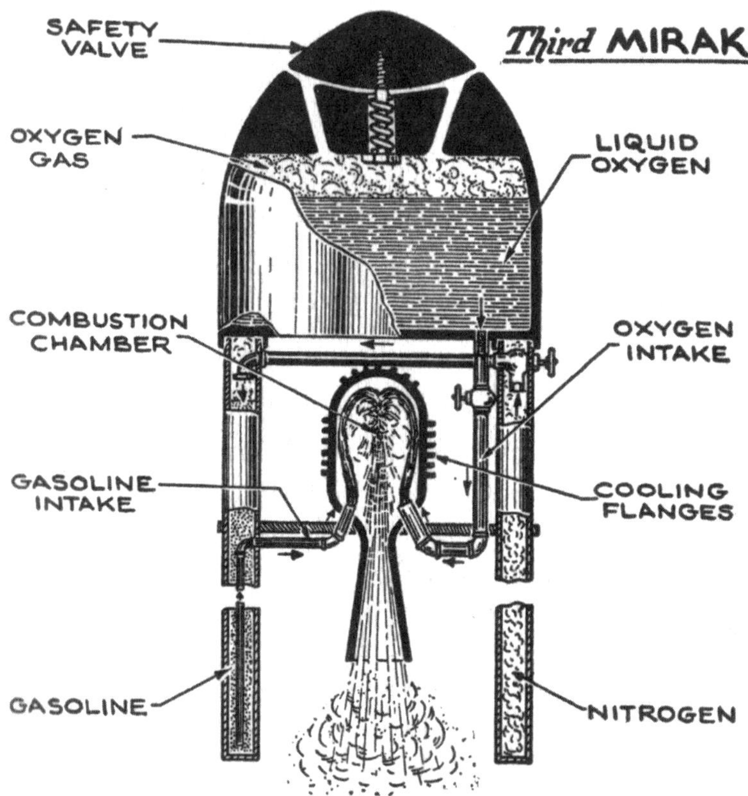

FIG. 2 SECTIONAL VIEW OF THE THIRD MIRAK (MINIMUM ROCKET) BUILT AT THE "RAKETENFLUGPLATZ," BERLIN

The flow of fuel to the combustion chamber is illustrated.

They built a series of Miraks (the German abbreviation for "minimum rockets"; rockets built on the smallest possible scale) to test out the action of various rocket designs. The story of the Miraks is pertinent here, for it contains the most complete experimental history of rockets that is available.

The engineers discovered first that to introduce fuel into the combustion chamber the inlets should be located near the throat of the nozzle, and not at the head of the chamber, as was previously the custom. It was found that when the earlier practice was followed the fuels were forced almost out of the chamber before they began to unite in combustion, thus wasting much of the available energy.

Considerable experimentation was devoted to the shape of the combustion chamber. The first Mirak, based on preconceived ideas, was cone-shaped, the apex being at the top. But this proved unsatisfactory, for the odd corners at the base appeared to obstruct the outflow of gases, and combustion was incomplete. It was clear that more room was needed at the top, where the actual combustion took place, to permit the gases to mix more freely. The chamber of the second Mirak, therefore, was cylindrical, and the third, built after exhaustive experimentation with all possible shapes, was that of an egg, as indicated in the illustration, with each end flattened off as a hemisphere.

This design was observed under test to create a curious but satisfactory effect. Whereas the pressure in the chamber during combustion is 200 to 280 pounds per square inch, the pressure necessary to force the fuels in was considerably less. It is believed that the currents of gas set up in the chamber during combustion produce an actual suction upon the incoming fuel. It is the theory of the Raketenflugplatz engineers that the fuels move upward along the walls, meet at the top, and are burned there.

It is not assumed that the egg-shaped chamber cannot be bettered. The shape of the chamber may actually depend in the last analysis upon the particular pressure created and upon the nature and combustion of the fuel. But it is evident that experimentation on sizes and shapes must continue until it is clear that the very best design has been discovered.

Not only is tremendous pressure created in the combustion chamber during the operation of the rocket, but also intense heat. In the combustion of liquid hydrogen and oxygen the temperature resulting is several thousand degrees Fahrenheit. The construction of the chamber walls, therefore, is a matter of prime importance.

The material must be capable not only of resisting the extremely high temperatures, but also of maintaining its strength. Otherwise the weakening of the chamber walls would cause the entire rocket to burst. The material used must be as light as possible, and must resist the corrosive effect of the rapid flow of heated, compressed gases.

The experiences with the Miraks are quite instructive on this problem. The combustion chamber of the first Mirak was of a heavy copper alloy, which was, however, insufficiently heat-resisting. The second Mirak was lined with steel, with an inner coating of a ceramic material made especially for the experiments. Although heat-resisting, the ceramic lining was lacking in strength, and as soon as the steel began to weaken the whole rocket exploded.

Research in light metals, such as aluminum, beryllium — which is not only light, but possessed of a melting-point as high as 3200° Fahrenheit — and molybdenum, which is heavy, but has a melting point of 4500°, should yield an alloy which provides not only lightness and strength, but also resistance to high temperatures. Metallurgy can offer to rocketry considerable aid in providing a metal adapted to the rocket's exacting needs.

Cooling systems can, in all probability, keep the chamber temperature within prescribed limits. The German engineers found that by adding flanges to the outer chamber walls they not only obtained a great radiating surface for the heat, but also materially strengthened the chamber.

It should be clear that the power developed by the rocket will rest, in the last analysis, upon the total amount of fuel the chamber will burn.

The size of the chamber is also a factor in the rocket's ultimate operation. Large chambers, as Dr. Goddard pointed out, are considerably more efficient than small ones. Although the heat and friction losses of a large chamber increase according to the square of the dimensions, its fuel capacity increases with the cube.

As larger rockets are built combustion chambers will be built larger for efficiency. Perhaps a number of them, interconnected by tubes to maintain equal pressures, will be necessary. The latter device, while increasing the total weight of the chambers, will probably consume a higher net amount of fuel before the chamber weakens from over-heating.

IV

The problem of the exhaust nozzle is likewise intimately associated with the final success of the rocket. Experience with gas turbines, which operate upon the reaction principle, as well as with rockets, has shown that the reaction developed depends upon the shape and length of the nozzle. For the nozzle not only increases the pressure of the gases in the chamber (by the constriction at the throat), but also utilizes the energy released by the gases as they expand on their way to the outside.

Dr. Goddard found that the size of the nozzle would depend upon the medium (air or vacuum) in which it was being used, and also on the velocity of the gas. If, for example, a nozzle is too long, or flares too much for a high velocity of ejection, the gases will not be able to expand sufficiently in passing along the nozzle, and separate currents will be set up that will cut down the velocity.

The experience of the Mirak-builders convinced them that the higher the pressure in the chamber the longer the ejection nozzle must be. Naturally in the small experimental motors the mathematically correct nozzle was not found to be necessary, though exact calculations have been made. The exact design of the nozzle will evidently vary for the type of fuel, the pressure in the chamber, and the ejection velocity.

In the construction of the nozzle, as well as of the combustion chamber proper, the problem of heat resistance demands attention. It has been found in some cases that during the entire combustion the greatest heat was developed either at the throat of the nozzle or in the nozzle itself. It may be necessary, therefore, that the construction of the nozzle be more durable even than that of the combustion chamber.

In this connection an investigator, Mr. Clyde Fitch[1] states that:

> 'The question of refrigeration or cooling has been answered by gas-turbine experimenters. For greatest efficiency the expansion of the gazes [through the nozzle] should take place adiabatically — that is, without loss of heat. Should heat be lost while the gases are expanding through the nozzle the velocity will be reduced. As a subterfuge for cooling — without loss of heat — a waterspray is injected into the exhaust nozzle. The water immediately evaporates and absorbs large quantities of heat, thereby cooling the entire system. This process at the same time expands the water along with the gas, and the heat absorbed by the water is converted into useful energy.'

This ingenious arrangement, serving so many ends at once, may well yield to rocket experimenters a clue to the solution of their own vexing heat problems.

The location of the combustion chamber and exhaust nozzles in the rocket are also complicating elements in rocket design. For greatest stability of the rocket in flight the reaction should be developed ahead of the centre of gravity. In other words, half the weight of the rocket must be behind the exhausts.

In the designs of practically all the early rockets this principle was disregarded, because it was feared that the flaming exhausts would heat excessively whatever was even remotely in their path. And in view of the fact that the greater part of the rocket weight would be composed of fuel, it would be necessary, in effect, to put part or all of the highly explosive fuel in the indirect path of the exhausts to fulfill the requirements for stability.

1 Clyde Fitch, report to the American Interplanetary Society, January 16, 1931.

The design of the third Mirak, with several legs, one of which contains the petrol for combustion, is an approach to the solution of this problem, and an avoidance of the dangers mentioned. In the third Mirak the combustion chamber is certainly close to the centre of gravity of the entire rocket. Future designs may see the combustion chamber near the nose of the rocket, with a ring of exhausts, circling the rocket, shooting their flaming gases away at a safe angle.

V

The fuel chambers, the second part of the rocket, at the start of a flight will probably hold the greater part of the total weight of the rocket.

The chambers must be placed so as to ensure the stability of the rocket in flight, whether they are empty or full. They must be built of the lightest possible material, yet be sufficiently strong to contain the fuels, and they must be impervious to corrosion by the fuels — dry or liquid. Finally, they must be equipped with means to pump the fuels into the combustion chamber.

It is obvious that if the fuel chambers are badly placed the unevenly distributed weight will cause the rocket to be unstable in flight. Its motion, furthermore, will admit of no accurate control.

Since the weight of the fuel is such a great percentage of the total two methods of locating the chambers are possible. They can be centered on the vertical axis of the rocket, or round the circumference. In either case the lowering of the fuel level will cause no shifting of the centre of gravity from the axis. To avoid the change of the centre of gravity in a plane perpendicular to the axis it may be well to place one fuel ahead and the other behind the centre of gravity.

In the third Mirak the oxygen chamber was placed on the vertical axis and in the upper part of the rocket. The chamber for petrol, the second component of the fuel, was in a metal leg — designed like the stick on the skyrocket — to aid the stability of the rocket in flight. The weight of this leg, however, was balanced by a second leg, holding nitrogen gas under pressure, the purpose being to force the petrol into the combustion chamber. An increase in the number of legs to four or eight, distributing them round the circumference of the rocket, would further increase the stability of the weight.

Since the fuel chambers must occupy such a large part of the rocket the chamber walls will naturally contribute greatly to the weight of the projectile. The material of which they are constructed must therefore be as light as possible. During the acceleration and deceleration of the rocket the heavy fuels will press against the chamber walls with considerable force. The walls must therefore be strongly reinforced. When liquefied gases, such as liquid oxygen and liquid hydrogen, are used as the fuels the chamber walls must maintain their strength at extremely low temperatures.

The last condition is probably the most exacting in the construction of the fuel chamber, and may be the governing factor. Due to the brittleness of ordinary materials, such as steel, at very low temperatures, new materials must be found. For the oxygen container, Professor Oberth suggests the use of copper with the addition of a little zinc. Copper maintains its strength at low temperatures. and even at -182° Centigrade (when oxygen liquefies) it has a tensile strength of 42,500 pounds per square inch. Duralumin, an aluminum alloy, because of its lightness and strength, is particularly suited for the oxygen container, and has been used extensively by German experimenters.

For a hydrogen container lead is recommended by Oberth. Lead alloyed with copper has a tensile strength of 64,500 pounds per square inch at -253° Centigrade (the temperature of liquid hydrogen). For fuels used in place of liquid hydrogen, such as petrol, alcohol, or the hydrocarbons, the question of low temperature strength is not so important, and duralumin will probably be satisfactory. In these fuels the more important requirement is general strength as well as resistance to corrosion.

In the use of liquid oxygen and hydrogen the fuel chambers must be heat insulated, for above -182° oxygen vaporizes rapidly, and above -253° hydrogen becomes a gas. The fuel will be located only a short distance at best from the combustion chamber where the temperature will approach 2500°. Double walls, making the fuel chambers effective thermos vessels, must be used. In large rockets some additional cooling system may be necessary, especially if the fuel chambers are distributed round the circumference of the rocket. Then the friction of the passage through the air may be sufficient to heat the fuels so that they will boil furiously.

Despite all cooling precautions it is naturally impossible in practice to maintain the low temperature necessary to keep oxygen and hydrogen entirely liquid. Vaporization must be expected, and, since the accumulating gases exert an ever-growing pressure in the fuel chamber, safety-valves must be provided to keep the pressure within limits.

Again we turn to the German experiments for the lessons learned concerning this problem.

The first Mirak had no safety device, and the oxygen chamber exploded incontinently. In the second and third Miraks, however, the oxygen chambers were equipped with simple safety-valves, the pressure in the third Mirak being regulated at 90 pounds per square inch.

The problem of pumping the fuels into the combustion chamber against the pressure in the chamber has engaged the attention of practically all the leading rocket experimenters. So serious is it to Robert Esnault-Pelterie that he stated recently to the writer that his first task in the design of a large rocket for terrestrial or interplanetary voyaging would be to spend several years in the construction and testing of pumps.

Hermann Oberth likewise believes it to be of importance, sufficient to engage considerable attention in his book *The Rocket for Space Travel*, and he has provided for elaborate pumping systems in his rocket designs.

Both men emphasize the fact that rockets will burn a comparatively short time and the fuel must be supplied in large quantities and in determined proportions. The relative weights of oxygen and hydrogen, or oxygen and alcohol, required for perfect combustion are accurately known to chemists, and they must be supplied unfailingly. The failure of one of the fuels to stream into the combustion chamber in sufficient quantities may well mean disaster to the vehicle.

The pressures that will be developed in large rockets will range from 300 to 700 pounds per square inch. Turbine pumps can develop pressures of only 280 to 320 pounds per square inch, and their use for rockets is thus limited. With reciprocating-pumps pressures up to 1100 pounds per square inch are possible, and they might be adapted to the rocket's need.

The builders of the Miraks have solved the problem in their miniatures without external pumps. They have utilized the pressure of the vaporized oxygen above the liquid level to force the liquid into the combustion chamber; and for the pumping of petrol they have used nitrogen gas under pressure.

The apparent success of this arrangement has been obtained, as was stated, by the curious suction effect upon the fuels caused by the shape of the combustion chamber. The oxygen is under a maximum pressure of 90 pounds per square inch, and the nitrogen exerts 150 to 180 pounds per square inch upon the petrol to force it into the combustion chamber against a pressure of 280 pounds per square inch.

The art of rocketry will have received a great impetus if this simple and efficient device will obviate the need of external pumps. For pumps will undoubtedly create considerable additional weight and complications in the entire rocket design.

VI

The capacity of the third of the fundamental parts of the rocket, the pay-load compartment, is the final criterion of the success or failure of the entire rocket. Although the ability of a rocket to ascend to a new and hitherto unreached height may be theoretically imposing, the value of the rocket will depend upon what apparatus, instruments, passengers, or freight can be carried. If the propulsive force equals only the weight of the rocket without pay load, then for all practical purposes the rocket is a failure.

In the skyrocket the pay load is a charge of colored stars, released during the flight; in lifesaving rockets it is a coil of rope; in the meteorological rocket it consists of instruments for recording the nature of the upper air and of space, as well as parachutes or other landing devices; in rocket planes it will consist of human beings, the control equipment, and the mail or freight to be transported.

The pay load even in the best designs will be only a small proportion of the total weight and can be placed safely near the nose. This will aid in the navigation of the craft if it is to hold human beings, and in the work of observation in the meteorological rocket. Naturally, if the landing device is to be a parachute which opens at the peak of the rocket's flight and permits it to descend gently to the earth, it must be placed in the nose. The pay load must also be located in such a place, to be free from the extremes of temperature found in the other parts of the rocket, and out of the possible path of exhaust gases.

The question of the actual distribution of pay load will be the last of the problems to be settled. It will be axiomatic in all rocket designs that the margin of weight left for pay load should be as large as possible.

The Miraks were not equipped to carry pay load. They were simply test rockets, it being understood, however, that a pay load could be encased in the nose. The efficiency of the design of the third Mirak may be indicated by the fact that despite the small size of the combustion chamber — it is no larger than a hen's egg — it developed an upward impulse of eighteen to twenty pounds. The entire rocket loaded with fuel weighs only nine pounds, the accelerating force being the difference of about nine pounds. Loaded with only a litre of oxygen and a half-litre of petrol, such a rocket could be accelerated upward 32 feet per second per second at the start; and after the weight has been decreased by the ejection of part of the fuel the acceleration might be as high as 64 feet per second per second. Burning for 32 seconds, this fuel would be sufficient to send the rocket three miles into the air, although air resistance might reduce the height to two miles.

Naturally, if built upon a larger scale, the Mirak's propulsive force would increase much faster than the additional weight, and the achievement of an altitude of fifteen to twenty miles with a similar rocket, only three or four times as large, should be possible. These figures are given to indicate the power developed by extremely small rocket engines.

For a safe landing on the earth after its flight the rocket must be equipped with devices to check its earthward plunge. The use of wings, which could act also as stabilizing fins, is naturally suggested by the example of the aeroplane. The wings must be light, yet strong and heat-resisting, for they may cut through the air at speeds of five hundred miles an hour and upward. Wings, however, are more useful in permitting a glide to earth than in actually braking the speed of a descending rocket. The use of a parachute released from the nose is of more value. Parachutes are relatively light, easily compressed into a small space, and yet can support heavy weights in the air.

A parachute may be enclosed in the very nose of the rocket, under a spring set to open and release it. When in the rocket's ascent the pressure of the air upon the nose becomes less than the force of the spring the parachute is released; and when the downward plunge begins it is opened by the rushing air. Other methods for release may doubtless be used, such as clockwork devices, wireless control from the earth, photoelectric cells, etc.

An adaptation of the autogyro to the rocket was made recently in the flight of the rockets of Reinhold Tiling, the German engineer. Small wings, capable of rotating the entire rocket shell, thus giving an autogyro effect, were used by Tiling, the wings unfolding automatically when the rocket began to descend. The experiment, according to observers, was a complete success, and may open the way to a new approach to the landing problem.

If rockets are to be shot high above the earth or to other worlds the problem of landing without the cue of excessive fuel will assume major importance. A rocket sent to the moon, for example, will probably return to earth at the speed of seven miles per second, with which it departed. Unless as much fuel is used to check the speed of descent as to start it an adequate means of utilizing the braking power of the air must be found. The parachute, by all present evidence, seems to be the answer.

VII

The rocket shell, the fourth of the rocket's parts will bear the brunt of the strain and friction of flight. It must be built as lightly as possible, yet must be able to stand the enormous pressure of the swift movement through the air, and the heating due to friction, and the unrelieved glare of the sun in interplanetary space. The shell must be capable of resisting sudden changes of temperature, as well as stresses and strains. Aluminum alloys such as duralumin are favored by rocket experimenters, and further research with them may be expected to yield increasingly better adaptation to the exacting requirements of rockets.

As a projectile that is to travel through the air at high speeds, the rocket must be of such shape as to reduce the resistance to a minimum. It must be stable in flight, and its course must be controlled either by a navigator on board or by mechanical steering devices.

The science of ballistics is of great aid to rocket engineers in studying this phase of the problem.

The experience of at least a century of observation of projectiles in flight, and the recording of considerable test data, has indicated that the resistance encountered in flight depends for a given speed upon the shape of the projectile, principally upon its cross-section.[1]

There is much experimental data on projectiles up to sixteen inches in diameter, and with muzzle velocities up to 3000 feet per second. The study of such data, however, reveals discrepancies in the experience of various investigators. We have come to believe that the resistance of the air increases as the square of the speed, but this is found to be only roughly true. At low velocities it is definitely untrue. It will serve, however, as a basis for study until actual experimental data with rockets can build up a new science of ballistics.

Rockets of varying shapes must be sent aloft, and the total resistance of the air must be measured, so that the shape providing the least resistance may be discovered.

Experience with stability of projectiles has shown that:

> all are unstable along the axis of flight — that is, the axis tends to rotate in the plane which is perpendicular to the line of flight. Small weight differences pull the rocket out of balance, and inequalities in the propulsion force at the tail exert a powerful leverage to rotate still further the projectile or rocket.[2]

The rifling of bullets to give them a rotary gyroscopic motion and thus to provide stability, is hardly adaptable to the rocket, and it would seem to be necessary to

1 Long thin rockets, although less stable in flight, present less of a cross section to the air, and encounter, in general; less air resistance.

2 Fletcher Pratt, report to the American Interplanetary Society, December 19, 1930.

provide a gyroscope actually within the rocket, with its centre on the axis of flight. The gyroscopic action would work powerfully to prevent the wobbling of the axis. The placing of the combustion chamber ahead of the centre of gravity, as mentioned previously, would also add to the stability.

Hermann Oberth suggests a gyroscope connected through some electrical device with the exhaust stream, and arranged to alter the direction of the exhaust, and therefore of the propulsive force, when the rocket gets off its course. The action of such a device would have to be exceedingly delicate and its control instantaneous, otherwise it would defeat its purpose, and, in fact, increase the wobble of the ship.

The question of steering depends naturally upon whether the rocket is to hold human beings, or whether it is to be sent upon a directed flight from earth, or controlled from earth. In any event the actual steering must utilize the rocket exhausts. Tubes will be placed either along the side of the shell, or be honeycombed at the tail. In either case, to turn to the right, exhausts on the left will be used; and to turn to the left the exhausts on the right.

To brake the rocket's flight exhausts at the nose have been suggested, but since it is likely that while operating in air the rocket would find itself bathed in its own flaming gases by this procedure, it may have to be abandoned, although it should work in airless space. For general braking by the use of fuel the rocket must be turned end for end and go forward tail first.

To shoot an unmanned rocket on a predetermined course with any hope of accuracy will be very difficult, for it has been discovered that even a rifle held rigidly in clamps and firing identical bullets at a target a hundred yards away will not drive them through the same hole. Any deviations in the course of a rocket expected to travel considerable distances will be magnified enormously during the flight.

Control of the exhausts by wireless from the earth is of course possible, though the science of wireless control is in its infancy. The question of steering is naturally bound up intimately with that of stability, for no possible steering device can operate unless the rocket's flight is sensitive enough to obey it.

The available data on the rocket in flight is, perhaps, the most meagre in the whole problem, and ambitious uses for the rocket must await the growth of a science of rocket flight. The construction of the equivalent of aeroplane wind tunnels should aid immensely, and wind-tunnel tests can give indications to rocket-builders how their projectiles will function in practice.

VIII

Since the rocket must carry its own fuel and provide extensive fuel compartments on a long journey, as an ascension into interplanetary space or to another world, a great deal of the rocket weight is a useless burden after the fuel has been partly consumed. To reduce this dead weight the step rocket, in various forms, has been suggested by many experimenters.

Dr. Goddard has patented several such devices, the principle of all being the division of the total rocket into a number of smaller rockets, in the order of railway coaches. Each step, as it is called, is a complete rocket motor, containing fuel, combustion chambers, exhausts, etc.

When the rocket starts on its flight only the motor of the first and largest step is operated. It lifts the rocket until the fuel of the step is exhausted, then it is detached from the rocket by a controlling mechanism, and either exploded into small bits or allowed to float back to earth by a parachute. The useless weight is thus eliminated. The next step is then set in operation, and it too is detached when its fuel has been consumed. Finally only the nose of the rocket will remain, holding the pay load and a rocket motor for the return to earth.

The step principle is one of the most valuable that has been developed in rocketry, and considerable investigation by experts in America and abroad has shown that the step rocket will ascend to a far greater height than a unit rocket of the same weight.

IX

In this brief review it is impossible to do more than to illustrate the complexity of the rocket problem, and to point out the necessity for the evolving of a complete science of rocketry. Upon the foundation of the principle of reaction a dozen sciences must be employed to construct a vehicle to travel faster and to go higher than man has hitherto attempted.

The rocket's needs are undoubtedly peculiar, and many of them seemingly incompatible. Only extensive tests will determine how the hundred and one elements can be composed to form a messenger of space. The rocket is, as Dr. Goddard has stated, the most efficient heat engine known. It must be tamed, controlled, and utilized. To this end experimenters in half a dozen nations are now working with determination. It is inconceivable that with so much energy, devotion, and technical skill they can do anything but succeed.

CHAPTER VII

HARNESSING THE ROCKET

I

EXPERIMENTS on the rocket have demonstrated one thing conclusively — that it is a means of propulsion for vehicles travelling only at extremely high speeds.

The reason for this may be found in the nature of the machine. For the obtaining of the greatest efficiency from a rocket — i.e. extracting the most propulsive force from the burning of the fuel — the vehicle must travel at approximately the speed of the expelled gases.[1]

Furthermore, the higher the speed of expulsion of the gases the more efficient the fuel.

Dr. Goddard found in his 1919 experiments that speeds of gas expulsion from two to 3000 feet per second yielded 15 to 20 per cent of the fuel's energy — but if the speed of expulsion were increased to 6000 or 8000 feet per second the efficiency rose to 60 or 70 percent.

By these tokens rockets must travel at speeds of at least one mile a second for a practical expenditure of their fuels. And since we have seen that the rocket must lift all its fuel, it can succeed in reaching the upper atmosphere, or interplanetary space, only if the fuel energy is utilized to the maximum.

Recent studies have shown that a rocket travelling at 25 miles an hour utilizes but 1.2 per cent of the fuel's energy while at 65 miles an hour the efficiency would be 2.8 percent; and even at 350 miles an hour only 13 percent of the possible fuel energy can be obtained. All land and water vehicles, and aeroplanes to operate in low altitude, are thus ruled out of the future of rocket propulsion — for the air resistance near the earth's surface makes speeds higher than 350 miles an hour virtually impossible.

1 This may be explained as follows: The obtaining of energy from rocket fuel consists in burning it, causing it to release its heat, which in secondary translation of the energy into the pressure of the resulting gases, and then a transmission of this pressure into kinetic energy of the gases so they are exhausted. The utilization of the gases to the utmost means that as much of their energy as possible must be given up to the rocket. In other words if the gases leave the target with no kinetic energy left they have given it completely to the rocket's propulsion. Expressed mathematically, the gases as they leave the exhaust must have zero velocity with relation to the rocket. The rocket for this condition of maximum efficiency must travel forward at the same speed that the gases are being ejected backwards.

With these conditions outlined, what field of usefulness has the rocket where it may utilize to the utmost its peculiar gifts?

To this question the rocket is affording its own answer; and, theoretically at least, it has a future that will justify its development even if a space flight is never made. As a result of the experiments of the past three years the rocket is creating a significant place for itself in our civilization. This future of the rocket as an instrument for examining our upper air, as a new weapon in warfare, and as a vehicle capable of flying in the stratosphere, deserves attention.

II

The future of the aeroplane for long-distant flights, according to such authorities as Colonel Lindbergh, Dr. Goddard, Max Valier, and others, lies in the utilization of the stratosphere, or the upper and thinner layers of the atmosphere. Not only will the 'plane then be above storm-levels and free from the menace of fog, ice, rain, and snow, but it will also be free of the devastating air resistance that eats up the 'plane's power.

Flying at ten miles or more above the earth, and guided by instruments, free from danger of storms and air resistance, the 'plane can acquire speeds of 500 miles an hour and more, to cross the Atlantic between noon and sunset!

Surely a fertile field for imaginative contemplation!

We have shown, however, that the propeller aeroplane is a child of the atmosphere, and the use of the rarefied layers to permit higher speeds is equivalent to removing a ship from the ocean in order to eliminate the resistance of the water to its passage. There is no doubt but that aeroplanes will be built to fly at much higher levels of the atmosphere than heretofore, with equipment to provide sufficient air for support of and traction for the motor. In fact, an announcement by the Junkers Aeroplane Company in the spring of 1931 indicates that a new type of aeroplane, to fly through the stratosphere at 500 miles an hour, is actually under construction in Germany.

But the propeller 'plane is not naturally fitted for stratosphere flight. What is needed is a means of propulsion that does not depend on air as a medium, and, better, one that operates most perfectly and efficiently in the absence of air. A perfected form of this instrument has already been found in the rocket.

Writing as early as 1921, Goddard predicted transatlantic service by rocket-planes for the carrying of mail and valuable freight, at speeds of 500 miles per hour and upward. The problem was, even then, according to this expert, theoretically possible of solution. More recently a patent has been granted to Dr. Goddard for a combination propeller and rocket-driven 'plane that will travel at high altitude and at high speeds.

Max Valier and Hermann Oberth both planned the construction of rocket-planes that would rise from Berlin, ascend thirty miles into the air, and, describing a graceful curve through the upper stratosphere, descend gently at New York one hour later.

To Esnault-Pelterie a Paris-New York flight of twenty minutes by means of rockets appears well within reason.

Such 'planes, using the extremely rarefied layers of the air, could achieve speeds up to 3000 miles an hour.[1] For at ten miles above the earth the air density is only one-tenth that at the earth's surface, and of thirty miles it is not over one-thousandth that at sea-level. The extreme tenuity of the air between these two levels eliminates the only bar to high speeds — that of air resistance.[2]

III

The transatlantic rocket-plane will probably follow years of experimentation with small torpedo-shaped rockets sent from one point to another under automatic control, and from one city to another carrying mail. The rocket passenger 'plane will then be a logical development.

Since it is to travel at altitude where the air is too tenuous for breathing, and where a temperature as low as -75° Fahrenheit prevails, the rocket-plane must be sealed tightly while in flight, and have equipment to supply oxygen as well as heat for its passengers.

The vehicle would probably be shaped like a torpedo, carefully streamlined, and possessed of wings to sustain it ascending and landing, but which could be folded or collapsed into the 'plane during the major part of the flight.

1Valier suggested that the greater part of the flight be at a height of thirty miles, while others believe that ten miles is sufficient.

2 Air resistance increased as the square of the speed of a projectile, or air vehicle. If the air density at an altitude of ten miles is only one-tenth as much as at sea-level, a vehicle could travel more than three times as fast with the same total resistance. Speeds, therefore, of 500 miles an hour and upward at a height of ten miles are distinctly practicable. At thirty mile's, where the density's only one thousandth that at sea-level, the normal speed could be increased thirty times, making 5000 miles an hour a possibility.

In the belief of Mr. Harold A. Danne, an American aeronautical engineer, the 'plane must be equipped to land upon water if the need arises. And since the water is often a rough place upon which to remain floating the size of the 'plane must be large enough to ensure marine stability. This would mean a ship at least 170 feet long and 30 feet in diameter.

In the flight planned by Max Valier the rocket-plane, with its crew, passengers, and cargo near the nose, and the fuel and rocket exhausts at the tail, would leave Berlin and shoot upward at an angle of 10°. Its speed would increase gradually as it rose higher and higher; and the angle of its flight slowly incline to 45°. The plane would quickly pass through the lower air-levels and reach the thirty-mile altitude, the ascension taking about fifteen minutes, the 'plane having travelled meanwhile about 250 miles.

Below, the details of the earth would be lost in a blur, though the horizon would be five hundred miles away. More than 750,000 square miles, one fourth the area of Europe, would be visible at once to the eyes of the 'plane's passengers.

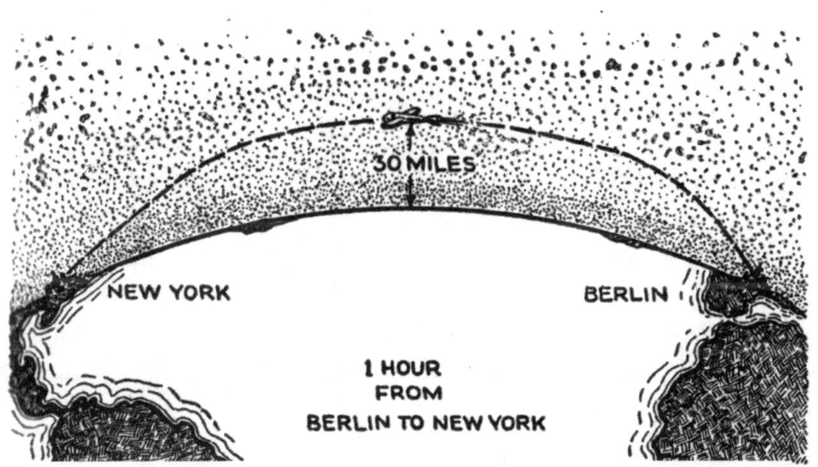

FIG. 3 THE FLIGHT OF A ROCKET-PLANE ACROSS THE ATLANTIC THROUGH THE EXTREMELY ATTENUATED AIR AT A THIRTY-MILE ALTITUDE

Overhead the stars would be visible as the 'plane, its course now levelled parallel to the earth's surface, shoots ahead at nearly four thousand miles an hour. Little power would be needed at this stage of the journey, for the craft would have the impulse of a bullet shot from a gun to maintain its flight. With gun-sights trained upon the stars the navigator would plot the course of the craft across the Atlantic, while the passengers, comfortable in the heated cabins, would have before them the unparalleled display of the aurora rising at the poles. Below the earth would seem thousands of miles away.

For only three-quarters of an hour would this part of the flight endure. Swiftly the Atlantic would be crossed. Europe would have disappeared beyond the horizon, and the outlines of the North American coasts begun to emerge in the west. Then slowly, as the telescoped wings are spread out to pierce the thin air, the 'plane would be nosed down. Its flight would become an elongated glide as it rushes into the denser air-levels. The earth would shoot upward to meet the 'plane, and the last part of the course would be adjusted for a careful landing at the rocket-port, one hour and a quarter after the start.

The use of rocket-planes to revolutionize our long distance transportation will probably precede by many years the construction of a space-ship. For the same problems of fuel and design that beset the astronauts hinder the construction of the 'plane. The difference, however, is that while a space flight with present fuels is hardly practical, the rocket-plane can really be built, though its operation might at present be uneconomic.

Valier has calculated that with fuels providing expulsion speeds of 7200 feet per second, 75 percent of the total load of his transatlantic rocket-plane must be fuel, and of the 25 percent remaining not more than 15 percent could be devoted to the pay load — the weight of the operator, equipment, passengers, or freight. Fifty-two tons of fuel would be necessary, in effect, to transport one ton of net pay load from Berlin to New York. If new fuels could be developed and controlled to give an expulsion of speed of 13,000 feet per second, the ratio of fuel to pay load could be reduced to sixteen tons to one. But even this ratio hardly provides a basis for economic operation, except for valuable parcels. Such rocket-ships would have to charge an excessively high rate for passenger transportation.

The final solution of the problem of rocket-plane transportation at high altitude and high speeds is not yet here, but by all evidence it is not far away. Scientists and technicians generally are willing to consider it seriously, even though some of them do not admit of the possibility of a moon flight. The rocket-plane certainly presents an opportunity for relatively early use and practical benefit.

The serious attention of men of science cannot but help to extend the scope of investigation and research and to lead ultimately to the interesting of capital and the best technical skill in the building of a rocket-plane. The attitude of French scientific authorities to the startling proposals of Esnault-Pelterie is perhaps indicative of the new trend. Louis Forest, writing in Le Main of an address by M.

Pelterie, said:

> I attended this talk and afterwards heard Rodolphe Soreau, President of the Society of Civil Engineers of France, and Messieurs Urbain and Sichot, both members of the Institute [de France], calmly discussing the passage of a ship to New York in twenty minutes and a trip to the moon. There should be no mistake about it. It a not a question of Jules Verne's ideas and fancies, but a question of exact conceptions. M. Pelteries's book *L'Astronautique* is not a child's plaything, but is crammed with scientific calculations and precise formulae.

Signs of the coming of this transatlantic rocket-plane service will not be wanting.

With growing control over liquid fuels ambitious experimenters will equip aeroplanes with rocket motors and make flights to greater and greater heights. Rocket-propelled gliders will make their appearance and arouse the enthusiasm of inspired amateurs. The interest of 'plane manufacturers will be awakened by the widening field of the rocket, and they will open their laboratories and research facilities to intensive studies of rocket-plane construction.

Financial incentive for the first stratosphere flights will come at last, and from wealthy seekers of thrills. Men and women of means will offer considerable sums for the privilege of a flight to unattained heights, to view the earth so far beneath them, and to circle the globe faster than the sun.

IV

The rocket as the motive power for terrestrial or interplanetary flight has still to enlist the support of the great body of modern scientists and of financiers. As a mechanical instrument for studying the nature and phenomena of our upper atmosphere and of interplanetary space, however, the rocket already has the backing of capital and of a group of noted American scientists. The preliminary work toward the construction of such an instrument is proceeding with all the speed that ample funds and expert technical skill can provide.

The grant of $100,000 (£20,000) by the late Daniel Guggenheim to the Goddard experiments may be accepted as evidence of Guggenheim's belief that the rocket will become of definite scientific and practical value. With this fund, and assured of the assistance of a board of advisors,[1] Dr. Goddard has established a modern,

1 Including Colonel Lindbergh, Dr. R. A. Milliken, Dr. C. G. Abbot, head of the Smithsonian Institution, D. J. C. Merriam, President of the Carnegie Institution and Dr. W. S. Adams, Director of Mount Wilson Observatory.

completely equipped rocket laboratory at Roswell, New Mexico, where he plans to build the first meteorological rocket.

"Perfection of Dr. Goddard's rocket," said the statement announcing the grant:

> will mean that thermometers, barometers, electric measuring apparatus, air-traps for samples of air, and other instruments may be sent to extreme altitude to bring back such needed information.
>
> It is planned to shoot a rocket 50 to 250 miles and control its descent by parachute. Its greatest weight will be in liquid propellant, which naturally will be exhausted in flight, so that on descending the full-size rocket will weigh very little, considerably less than an aviator jumping with a parachute from an aeroplane. The advantages of the vertical ascent are obvious. Present methods of reaching high altitude are limited to balloons, whose ascent cannot be controlled, and whose place of landing is a matter of uncertainty. Recovery of such balloons depends upon the person who finds them. Quite often they are lost altogether, or their instruments tampered with to such an extent that they are useless.
>
> When one of the rockets is perfected — and it is admittedly a number of years away — much valuable information will be obtained in the fields of meteorology, astronomy, wireless broadcasting, aviation, and science in general. For instance, it is expected that a good deal will be learned about the spectrum of the sun above the ozone layer from 50 to 75 miles above the earth's surface.

Further knowledge will be gained of wireless broadcasting waves, which are known to follow the ceiling of the Kennelly-Heaviside layer. This is assumed to be a layer of air charged electrically. In the field of meteorology, and as a particular aid to aviation, it may be possible to prepare high-elevation weather-maps.

Such rockets, in the opinion of Dr. Goddard and his advisory scientific board, would provide an inexpensive means of obtaining a weather-map of high altitude, just as we now have them of the earth's surface.

This statement is apparently toned and qualified to shock conservative opinion as little as possible. Yet it must be admitted that the support of a financier of world-wide experience, and of men of science with reputations to guard, comes as a revelation to those who viewed the field of "rocketeering" as a visionary dream. Although to the practical reader the need for information of what happens hundreds of miles above us may seem at first academic, on consideration it cannot but take on the utmost importance.

The picture of the earth's place in the universe drawn in a previous chapter visualized the earth as a small spinning ball in a universe governed by multitudinous forces about whose nature we know little.

Born of the sun, the earth has always been subject to solar moods. But it is only recently that the extent of the influences that make us "children of the sun" has been discovered. We are born, we live and die, on a celestial speck of dust — subject to tremendous cosmic forces; and our destiny as a race is determined by these forces. Whether or not we know of them, whether or not we interest ourselves in them, they exist.

Many scientists in imaginative moments have pictured living worlds on the specks of dust we see floating in the sun-beams. Certainly such dust creatures would be amenable to our complete domination; yet no more so than we ourselves are dominated by the vast forces of the outer cosmos.

V

Toward the end of September 1907 part of the telephonic and telegraphic systems of the world were rendered useless, as though they had never existed. And other apparatus depending upon electricity was seriously affected. Astronomers, directing their telescopes at the sun on that day, noticed several dark patches, contrasting vividly with the grey-white molten surface. These markings were recognized as sun-spots, centers of solar activity — insignificant in the life of the sun. Yet upon the presence of these spots, 92,000,000 miles away, was placed the blame for the disruption of communication in our modern world!

Utterly powerless to alter the situation in the slightest, the people of the earth found that wireless reception during the years 1928-30 had altered noticeably for the worse, which caused attention to be directed to the fact that during the years 1923-26 it had been consistently good; and again sun-spots were blamed.

Recently we have been made aware of cosmic rays, waves of energy like wireless or light waves, that strike the earth. Although they emanate from far-off places, students of these mysterious penetrating radiations believe that they exercise a vital influence on our lives.

And the great electrical storms that sweep the earth; the periods of intense heat and equally intense cold; the long sieges of drought; the sudden eruptions of volcanoes and the equally violent earthquakes; the cyclones and onslaughts of insect and bacterial life — all are signs that we are creatures of the giant forces of nature.

Despite man's so-called progress he is the creature not only of his immediate environment, but also of the solar system and of the infinite spaces beyond.

He can no more afford to ignore what happens in the heavens than an ant colony can forget about the human foot that rests near it. A celestial disturbance can as casually wipe all life from the earth as a human foot could destroy the ants.

The Goddard meteorological rocket represents a determined attempt of man to understand and predict the ever-changing forces that control our destinies. When such rockets have been perfected they will be sent aloft to the upper limits and beyond the earth's atmosphere, and by mechanically operated scientific devices record just what is occurring in the sun. A science of solar as well as general cosmic activities, so far as they affect the earth, will gradually evolve, until the day comes, as we hope, when the vagaries of weather, temperature, and activity of wireless waves will not find us unprepared and helpless.

VI

It is quite possible that with the discovery of new fuels and methods of control the rocket will again become a dreaded weapon of war. Carrying their own means of propulsion, rockets could be sent high into the air, on flights of two to five hundred miles, and permit the bombardment of the heart of the enemy country from a safe distance.

A prediction of this was actually made by Professor Hermann Oberth, the German rocket experimenter, at a lecture before the Vienna Meteorological Institute. He visioned rockets equipped with cameras sent out to map enemy positions, and returning obediently to their own armies. By an extension of their range an army on one side of the earth could bombard an enemy on the other side with "a murderous rain of rockets containing poison-gas, which could exterminate an entire population." Herr Oberth himself admits carrying on experiments with rocket projectiles for war purposes, but only in the hope that by showing how horrible a weapon they could become they would act as an effective deterrent to warfare.

It is to be hoped that this use of the rocket, with its attendant horrors and destruction of non-combatants in war, will never be realized; and, rather than have the rocket become such a fearful agent of destruction, it would be better that it be forgotten altogether.

But the prospect of such methods of warfare may actually serve as a bar to the beginning of hostilities. If civilian populations were aware of the existence of an instrument able to devastate their homes, factories, schools, and public buildings, to destroy their children and lay waste all they have constructed, the great public opposition to war, which is the only effective Peace agent, might become operative. But the rocket exists, and its use in future warfare, both as a long-distance shell and as a new weapon for infantry, is a distinct possibility.

VII

From the point of view of actual accomplishment the rocket has not yet demonstrated its ability to escape from the earth. Many problems of fuel and its control, and of design and steering still face the zealous astronauts. Understanding of the peculiarities of the rocket in flight is still sketchy and incomplete, and public support still awaits more definite proof of the rocket's capabilities as a stratosphere 'plane and a messenger into the upper atmosphere.

In scientific circles there is undoubtedly much skepticism and not a little absolute denial that the rocket will ever conquer celestial spaces. And midway between that attitude and the optimism of Esnault-Pelterie that a moon flight can be made in twenty years are the predictions of men like Dr. John Q. Stewart, astronomer of Princeton University, that an interplanetary journey may be made a hundred years hence.

Even enthusiasts such as Esnault-Pelterie foresee years filled with progressively ambitious experiments before the terrestrial rocket flight or the space flight can be attempted. There will be years of trial and error not only with fuels, but with types and shapes of vehicles, with methods of providing heat and breathable atmosphere for passengers traversing the airless spaces. There must be protection devised against the possibility of collisions with meteors in space, against harmful rays from our sun and other suns, against physiological and psychological effects on the human organism of the reduced gravitational force in interplanetary space.

There will be heart-breaking failures, temporary losses of confidence by the public, attempts to legislate against the continuance of this "hare-brained scheme." The path of experimentation will be littered with the wreckage of machines and the bodies and dreams of men; and fortunes will be lost by those who believe that they have discovered the secret of space travel where others have failed.

The record of rocket experimentation of the near future will hardly be different from that of the past five years — with discouragement coming to many Opels and death to pioneering Valiers. Yet through the wreckage of dreams and machines there will be perceived gradual and determinate progress. Rockets will rise higher and higher into the atmosphere. Premature shots will be made at the moon, and the rockets will fall back to earth, either to crash there or to circle the globe as eternal satellites. All this will occur, and only when we are certain that we have the correct combination of fuel and equipment and knowledge of the outer spaces, and that we have as a result a machine obedient to our will, can the fast craft bearing passengers arise from the earth to plunge upward towed the moon.

PART II

THE FLIGHT INTO SPACE

CHAPTER VIII

THE SPACE-SHIP

I

THE space-ship that is to fly to the moon has been completed, and the International Interplanetary Commission views it with justifiable pride. Tonight at precisely nine-thirty the 150-foot metal monster, now waiting in its hangar, will go roaring into space.

Since the completion of the ship the construction camp in the Swiss Alps has been open for public inspection. A steady stream of political and social leaders, scientists, and educators have viewed this newest of creations. Day after day the hangar has been filled with thousands of people, pressing close to the railings that protect the ship, gazing at it with awe and amazement.

It is now apparent how wise and far-seeing were the great nations when they agreed to abandon the race to reach the moon, and pooled their knowledge, skill, and resources. For what had threatened to become a new source of national pride and jealousy had been turned into a joint endeavor, in which each nation could feel its contribution. Races unable to settle amicably their political and social differences had received from their scientists a striking lesson in international co-operation.

Into the construction of the 10,000-ton ship has been poured more than $100,000,000 (£20,000,000). This sum was appalling to those who had objected strenuously to the undertaking of the "mad project." But when it had been retorted that the cost was equivalent to that of only two battleships the critics were quieted. Supported by great public enthusiasm, the nations concerned with the construction enthusiastically gave their determined shares.

II

It is four-thirty in the afternoon of the day of days, and we wait with our baggage beside the path that leads from the hangar. Inside the hangar at our right the ship Terra is being charged with the fuel that will send men for the first time into the great spaces.

The day is clear and cold, typical of early March, and we shiver in our overcoats, wondering apprehensively how cold it will be in outer space.

Etched sharply against the white snow, guards vigilantly patrol the hangar and the steel rails over which Terra is to begin her flight.

The polished rails run off to our left. Inclining along a gentle slope for a quarter of a mile, they end abruptly where an Alpine valley begins. In the west, beyond the valley, height on height rise the jagged peaks of the Jungfrau and her sisters of the Oberland.

Three hundred yards beyond the path, pressed close to the outer fences of the construction camp, are innumerable thousands who already have waited hours for the great moment. Their voices and occasional shouts fill the mountains, echoing and reechoing. Perhaps they envy us, the eighteen fortunate ones whose names and histories have filled every newspaper on earth. More likely they are quite satisfied to be observers, and to permit us to make the experiment of this first voyage.

Beside us is our personal baggage, the accessories that we have tried desperately to reduce to the ten pounds allowed each of us.

As we wait impatiently for Terra to be drawn from her hangar we know that many unknown dangers must be faced in this first venture into the void. The unmanned rockets shot into space by landing on the moon demonstrated that the basic technical difficulties of the flight had been solved. Exhaustive tests on our ship, Terra, proved as far as possible its power to carry us to the moon and return us safely to earth.

But, despite every precaution, we realize that the journey may end in disaster. Some unsuspected weakness in the ship may be revealed, or we may fail in our duties at a critical moment. For each of the eighteen has his assigned task. He must be competent in an emergency to take part in the ship's operation.

It was of the utmost importance, therefore, to choose from the thousands of applicants those best fitted to make the first flight. Searching mental and physical tests had been given us by the Interplanetary Commission, and our skill and resourcefulness, our physical courage and character, had been examined pitilessly.

The six members of the crew, from the Chief, who is to command the expedition, to the master mechanic, the navigators and the cook, know the ship inch by inch. The Chief himself has superintended its construction from the first blue-print, and knows its every part intimately. Master of half a dozen sciences, and a leader of men, he is eminently fitted to cope with the terrible ordeal that faces us.

The great doors of the hangar open slowly. The giant space-ship, resting on its long, wheeled cradle, and propelled by huge cranes, emerges and moves over the rails toward us. The clamor from the crowd rises to that of a hurricane as the ship stops in front of us. The cranes are detached, and return to within the hangar. The ship, its shiny nose pointed toward the western peaks, is now in the position in which it will begin its flight.

In less than five hours its power will send the cradle shooting over the rails. At their end the cradle will be automatically released, and we will soar over the valley, upward and upward toward the airless space.

The tail, honeycombed with exhaust tubes, now points toward a desolate wall of rock beside the hangar. There must be nothing inflammable in the path of the incandescent gases that will shoot out when the journey begins.

The ship, with its sharp, streamlined nose pointed toward the setting sun, is like a great silver bullet forged by Titans. Or, with its three pairs of metal wings, it resembles some Gargantuan interplanetary beast, ready to return to its distant world.

As we stand in silence, gazing at this giant projectile that towers above us, we are swept by waves of anxiety and impatience.

All that we have learned of the ship and its power is forgotten in a moment, as we think with terror of the test it must meet. It seems impossible that this alien thing, with its smooth, gleaming bulk, can be made responsive to our wills. There is at the moment the rush of fright that one feels when looking down from a great height — the instinct to draw back before he is precipitated to a terrible death below. The vast realm of space seems at the moment a world we must not enter. The gamble with cosmic forces seems certain to end in disaster.

But attendants have placed a ladder against the ship's side, and at a nod from the Chief we climb one by one toward an opening near the nose. The Chief checks each man on his list — the five members of the crew, the ten scientific experts, and we two[1] who have been permitted to make the journey as representatives of the world's newspapers. As the shouts of the crowd rise to a thunderous bellow and a hundred newspaper cameramen photograph us from every angle our fears vanish in a glow of pride, and we step unhesitatingly through the doorway.

We find that we have entered a bare metal compartment, the air-lock, which is

1 The reader and the writer.

for use if we leave the ship in space, or wish to explore the moon. Passing through a second opening, we emerge into our living quarters, a small, rectangular, metal-walled room, the walls of which curve inward over our heads.

Hammocks are suspended from the walls; metal bookcases, completely enclosed, line one wall; lockers for our baggage are on another. Everything, we note, is rigidly clamped to the floor or walls. Two portholes, of a creamy, translucent glass, are in the curved wall. The glass has been specially prepared to soften the sun's blinding glare, and yet to permit entrance of its light and ultra-violet rays.

Suspended from ceiling and walls, and projecting from the floor, are hand-holds like those familiar to the strap-hangers of a past age.

Following our carefully rehearsed instructions, we stow our baggage in lockers and test the straps of the hammocks. The Chief has followed us up the ladder, and now observes our preparations. He has already scrupulously searched the ship for that terror of aerial explorers — the stowaway; for the presence of such an uninvited guest on this journey would not only burden the ship with unnecessary weight, but would place the stowaway himself in grave danger.

Twilight comes gradually as we complete the disposition of our baggage. We have nothing to do now but wait with impatience for the zero hour. No supper will be served, for we must have nothing pressing upon our stomachs at the start. In any case, we are too excited for food.

It is now after six o'clock, and through the still open doors of the air-lock we see Luna, a rather full, silvery crescent, rise slowly over the mountains in the south-east. As we look at her anxiously, trying to read in her features the destiny of our voyage, the Chief appears with the members of the crew. We step aside as they close and carefully seal the outer door. When they close the inner door the clamor of the crowd ceases abruptly, and a sudden silence falls on us. Within the double walls of the ship, which are separated by a vacuum, no sound from outside can reach us.

At once we realize that we are isolated from our familiar world. We are imprisoned now, and at the mercy of this metal monster. All of our earthbound instincts assert themselves, and a wave of panic sweeps us. The Chief, sensing our feelings, smiles at us reassuringly, and though his manner appears cool and confident, he betrays his nervousness. We play with him a game of pretended indifference to the dangers that face us.

At seven o'clock, with our nerves quieted somewhat, we reporters begin our first story for the Press. It is to be sent to earth in Morse code by flashes of light as soon as we have reached the critical speed of nearly seven miles a second, that will free us from the earth.

From our new positions as embryo explorers we describe its features — now, familiar to everyone — and write of our conflicting feelings as we wait for the moment of the start.

III

We write as follows:

The ship is divided into three parts: two auxiliary rockets that are to send the ship into space and accelerate it to a speed of seven miles a second; and the ship proper, consisting of our living-quarters, control and observation rooms, and the rocket motors that are to bring us back to the earth.

At nine-thirty, now two and a half hours away, the fast auxiliary rocket at the tail will start the ship over the rails, and, inclining its path upward through the air, accelerate it to three and a half miles a second. The empty hull will then be detached, and by an automatic explosion burst into thousands of fragments. The second auxiliary will then be operated, and when its fuel is exhausted it too will be released and exploded. The remainder of the ship, in which we now are, will then have the necessary speed to be free of the earth. The earth will be 1500 miles away.

The power will be shut off, and although our speed away from the earth will gradually lessen as the earth seeks to pull us back our momentum will be sufficient to carry us into the moon's field of attraction. The ship will then be drawn to the moon, toward which it will fall with increasing speed.

Delicate maneuvers will be necessary to check our speed, in order to make a safe landing on the satellite. It is our hope that the landing will be made forty-eight hours after the start. We will be permitted three days on the moon, and then the return to earth will begin.

The first auxiliary [we explain for our readers] is propelled by a mixture of pure alcohol and liquid oxygen.[1]

1 Although this fuel does not give so high an expulsion speed as a liquid hydrogen and oxygen mixture — the fuel of the other steps — this is a great disadvantage. In passing through the atmosphere a high speed will not be required until the alcohol and oxygen auxiliary has already been consumed. Alcohol was chosen because of its high specific gravity, and therefore a greater weight of it can be contained in a small space, with a resultant saving in the weight of containers.

The second auxiliary and the main rocket will be propelled by a mixture of liquid hydrogen and liquid oxygen, which will provide the dual advantages of high expulsion speed and high specific gravity.

To supply us with air to breathe and a normal earth pressure within the ship during its flight through airless space an artificial system of ventilation is used. Even now the air we breathe is being produced in our chemical tanks. To prevent this precious air from rushing out when we leave the ship for outer space, or to explore the moon, it is first necessary for one to go into the air-lock, re-closing the door to the ship proper. The air in the lock is then pumped back into the ship. Encased in space suits provided with oxygen apparatus, explorers can leave the ship through the outermost door.

The reverse procedure will be followed on returning to the ship. The door to the air-lock will be opened and then closed. Air will be pumped into the lock, and only when the same pressure as prevails within the ship is reached will the inner door be opened.

The provision for double walls for the ship was necessary to maintain a comfortable temperature in space. Otherwise, unprotected by any softening blanket of atmosphere, we will be exposed to merciless extremes. Facing the sun, we will have the full blast of its heat, in which no human can live. On the side of the ship away from the sun there will be only the pitiless cold of interstellar space. The double walls, encasing us in a virtual thermos flask, help to equalize these extremes to a comfortable level.

The hammocks [we continue] are covered with a soft, fluffy material, and laced with broad, felt-lined stops. Such apparent luxuries, we have been assured, are really necessities, and their need will be quite obvious when the journey has begun.

Supported from the wall by strong steel springs, the hammocks are to be our beds during the terrible minutes when the ship is acquiring the speed that will free it from the earth.

IV

To make the most efficient use of our fuel the ship must be accelerated as quickly as possible to its critical speed. The maximum rate of acceleration — that is, the rapidity with which our speed is increased — is limited both by the heating effect of air friction and by what our bodies can stand. For just as one is pressed to the floor of an upward-rising elevator as its speed is increasing, so will we be pressed to the floor of our accelerating space-ship.

The endurable limit of acceleration for a healthy man has been discovered to be 100 feet per second per second. The pressure of this acceleration we will feel as a fourfold increase in our weight. A man of 150 pounds will groan under a suddenly imposed load on his body of 450 pounds. To stand erect during such an experience being obviously impossible, we must be reclining in our hammocks until the critical speed is reached and the ship's power is shut off.[1]

We realize clearly, as we write, that life or death will be decided during the eight minutes of the acceleration. For if the acceleration is not limited to 100 feet per second per second the growing weight on us will become a vice that will crush us into a lifeless mass. And if we somehow survived we would find the ship speeding through space with fuel almost exhausted and no means to check our flight. Our ship would become a steel coffin, to end, perhaps, by a crash on some solar planet, or to fly on endlessly through the infinity of space.

Dwelling upon these horrible thoughts, it strikes us with startling force how much depends upon the skill of the crew during the first crucial minutes. The Chief and his assistants will themselves be lying in hammocks in the control room — the next room to our own — and weighted down by the crushing force of acceleration. In front of them will be the panel board of meters registering the vital data of the ship's operation.[2] At the side of the hammocks will project the fuel controls, six-inch levers, the slightest pressure on them stopping the flow of fuel into the motors. At their side, also, will be the controls that will uncouple the auxiliaries when the fuel is exhausted.

Against the deadening weight resting on them the crew must be able to lift their hands a few inches to control the levers.

Despite the fact that the controls have been placed as close as possible to the level of the hammocks, the exertion required for these simple acts will be enormous.

We have been writing steadily for over an hour, and when at last our story is completed we realize with a shock that it is a quarter of nine. Just three quarters of an hour to go!

1 An acceleration of "100 feet per second per second" means that at the end of the first second our speed will be 100 feet per second. One second later it will be 200 feet per second; at the end of the tenth second it will be 1000 feet per second and so on; until at the end of the 370th second we will be travelling at 37,000 feet a second, or approximately seven miles a second.

2 Its speed and acceleration, its direction and the temperature and pressure of the fuel combustion.

Anxious to see the place in which our life or death will be decided, we open the door to the control room. Six hammocks, two rows of three, are suspended from the walls. Lining the walls are portholes and numerous complicated-looking charts, indicating the path of the ship through space and the position it must occupy at each hour. These positions are to be checked by observations on the sun, moon, the earth, and the stars.

If the ship is discovered off the course rigidly determined for it the rocket motors must be started, and additional fuel used to regain our path.

A door leads to the galley and washrooms, and another to the observation tower near the ship's nose, and to the very nose, where there is a giant parachute to assist our landing upon return to the earth.

The banks of dials and meters fascinate us. Innocent, simple devices, they are the records of our destiny. Upon what they tell, upon the truthfulness of their records, this ambitious flight into the unknown will fail or succeed.

Our feelings are those of helplessness, a sense of littleness before the unyielding mechanical forces that will control us.

A warning word from the Chief tells us that it is time to strap ourselves into the hammocks. But fifteen minutes remain before the start.

We return to our quarters and join the group of scientists. We are constrained to sudden silence as we climb into the hammocks. Under the Chief's watchful eyes we tighten the straps so that we are held securely to our berths.

We sink luxuriously into the soft downiness, and envision for a moment the moon voyage in Defoe's Consolidator. "The person . . drinks a certain dozing draught . . . and dreaming all the way never awakens until he comes to his journey's end."

But the parting words of the Chief — "Good luck" — return us to reality. He leaves us, shutting the door to the control room.

On the wall beyond our heads the electric clock registers eighteen minutes after nine. Our bodies are tensed now, despite the warning to relax; and although we try to breathe normally our breath comes in irregular gasps, and we sense the quickening pounding of our pulses. We lie inert, gazing upward at the curved metal walls, trying to fix our minds on the future pleasures of the journey. But our minds are blank. There is perfect silence in the cabin.

At twenty-eight minutes past nine there comes a low, then an increasingly vibrant, hum from the ship's tail as the motors are started that are to charge the first rockets ....

Twenty-nine minutes past nine we hear a sudden roar. The ship seems to start forward. Then it stops. The rockets have been tested. Now we are ready.

CHAPTER IX

THE START FROM THE EARTH

I

As the half-hour approaches with studied slowness we can picture the Chief, as he lies on his hammock, gripping the lever that will plunge the ship from the earth. We cannot but feel with him in his terrible responsibility . . . eighteen lives and the fate of this beautiful winged creature at stake. Disaster or fame for himself at any twitch of his arm.

As the minute-hand covers the six instantaneously we hear the booming of a thousand cannons. The ship jerks violently. We feel a thrill of movement. Another jerk . . . and another. . . We have a momentary sense of lightness; and then the straps begin to tighten about us. Dizzily the room spins — the walls swing upward over our heads; the floor becomes a wall ....

The invisible hand of a giant comes to rest on our bodies, and begins to press . . . and press. . . . We breathe with difficulty.

The soft downiness underneath gradually hardens, to become a block of unyielding stone; our bones feel as though they are in a vice .... The roaring of the exhausts has become a howling fury, its rhythm striking us with physical blows ....

Nine-thirty-three the clock reads to our blurred eyes. Five minutes more of this — yet we can bear this giant no longer. There comes another jerk; steel bands are pressed into us by grinning devils, and they pull tighter, tearing into our flesh as we try to cry out. Our mouths open and snap shut . . . wordlessly . . . the monster on us presses . . . and presses . . . intolerably.

When consciousness returns the weight is gone . . . . Utterly exhausted . . . covered with perspiration . . . we groan in relief . . . . Our bones feel cracked, our faces are swollen, mouths dry. The roar of the exhausts has ceased. In utter silence dull hours pass. There is a faint stirring of bodies, and an uncertain voice, "We're in space now!"

The words mean nothing to us. We want only to lie here forever.

Although a long time has seemingly gone by the clock registers only fifteen minutes past ten. The electric lights burn hotly within the cabin, but at the portholes the eternal darkness remains.

A timid murmur of conversation begins, and then ceases. For as we lie here, still stunned, a new and terrifying sensation has gripped us. Through the confusion of our feelings comes the certainty that we are falling . . . down to the earth .... We grip the sides of the hammock frantically, for at any moment we expect the fatal crash. We try to cry out, but our lips open and shut. Our stomachs become constricted; violent waves of terror sweep us. . .

With horrible clarity we picture the earth rushing toward us . . . image after image of crashing disasters race across our minds. Our perceptions have become sharpened to fiery intensity . . . we must cry out!

Exhausted, we relax in the hammocks. In utter weariness the tension disappears. The sensation of falling remains. We grip the sides of the hammock, and repeat monotonously the formula: "We are not falling, we are safe . . . we are not falling…"

To aid us in this attempt at courage the Chief appears in the doorway. His face is pale and drawn. Walking unsteadily, yet smiling, he announces that the flight is successful. We are already four thousand miles above the earth. We notice that his feet are resting in the straps that project from the floor.

" Do not take any sudden steps when you get up," he warns us. "You are weightless now, and you need little energy to move about. Always use the hand-holds on the walls, floor, and ceiling."

With this assurance we gather our disordered forces and unbuckle the straps to rise.

But we move too quickly, and shoot rapidly toward the ceiling. Only outstretched hands save us from a collision. We push away from the roof, and with terrifying swiftness rush down to the floor.

Although we have expected it, this experience with weightlessness amazes us.

With the slightest pressure against the floor we fly into the air. We can float about through the room if we wish, or swim through the air. Curled up, we can fall asleep suspended on nothingness, between the floor and ceiling. We can lift any object now, no matter how large or how "heavy" it had been on earth.

But there are dangers that accompany the freedom weightlessness offers. We must restrain ourselves constantly, and exert only a small fraction of the energy we customarily use for normal actions. Otherwise we will find ourselves dashing violently against the walls or ceiling.

At the Chief's instructions we put on thick steel shoes. The floor has been magnetized, and we can walk about on it with some measure of safety and dignity. The shoes, which would weigh twenty pounds on earth, are lighter than air here, but they give us a sense of contact with a supporting surface. With new assurance we move to refresh ourselves after our gruelling experience.

II

But we again bow to the inflexibility of nature. Gravitation — a hindrance on earth — would now be welcomed. We cannot wash by filling a basin with water, for water in a weightless state would form into a globule, held together only by its surface tension. The act of washing would break the water into thousands of bubbles that would float about aimlessly through the room. We must squeeze the water from bags and saturate cloths with it.

At meals, too, during our journey, we will discover that we have begun a new life — for all liquids will have to be served in little bags, which we can squeeze to extract the contents. Solids will come in covered metal boxes. With sharp knives we will pick our food apart, spearing the pieces to convey them to the mouth.

Although we are still unaccustomed to the weightlessness of our bodies, we are eager to view the marvels of space. As soon as we are refreshed the Chief leads us to the observatory.

Crowded in the little room, we stand at the portholes that circle it and get our first sight of the depthless spaces. Like diamonds on black velvet, the stars crowd the immense field of space. Millions of times more numerous than they appear on earth, they spatter the depthless sky, shining coldly and proudly, without a twinkle.

But what holds our attention is the massive ball of earth at the right. Suspended in space against the black curtain of the sky, reflecting only the feeble light of the

THE GIANT ROCKET IN ITS HANGAR, READY FOR THE FLIGHT INTO SPACE *From "By Rocket to the Moon" courtesy UFA Films Inc.*

half-moon, she appears unreal. It is not more than a giant globe that we can reach out and touch.

Can it be that we are now ten thousand miles away from the earth, and increasing our distance five miles every second? We seem really to be motionless in the eternal silence. There is not an indication that we are rushing through space at a hundred times the speed of an aeroplane.

The earth is not, as we had imagined, beneath us, but is "at our side." We are apparently moving south-west across her surface. As on a great dark relief-map, we trace the western coast of Europe, the irregular blotch of Africa, the eastern coasts of the Americas, and between them, streaked with moonlight, the broad expanse of the Atlantic.

Brightly illuminated are the white patches of the Arctic wastes, and rising behind them a shimmering halo of glory, are the streamers of the aurora.

"Where is the sun?" we ask the Chief. We are now, he tells us, in the earth's shadow, its bulk hiding the sun from us. But in a few minutes we will escape from it, and we will witness the majestic spectacle of a dawn in space.

Meanwhile, as we look down, fascinated, we detect the stately turning of the earth. Eastern Europe dips slowly out of sight, and the Americas turn their faces toward us. In a few hours continents will have become irregular mud-pies set in small puddles, great rivers will be watery cracks, mountain chains minute corrugations. In this compressed map of an earth cities and countries will have lost their identity.

As we exchange glances in the observatory there is becoming more evident among us an increasingly acute homesickness. We feel a vague loneliness despite the presence of our fellows, and sudden inexplicable tendencies to tears.

The communications operator appears to report a message in light signals from earth. The Vienna Observatory has sighted us, and the news has been flashed everywhere to a jubilant world.

It is our chance to send back our story. Leaving this tower of wonders reluctantly, we descend carefully with the operator. We hand him our story, and hurriedly return for the dawn in space.

Just as we reach the observatory we hear the exclamations of wonder from the men at the portholes. We take our places at ports on the right, and see at once that we have been turning steadily away from the earth. It seems to be receding toward our tail and rising above us.

The earth's western limb, showing the Americas and the broad Pacific, is illuminated faintly by the crescent moon. But, as we watch, we see circling the western arc a snakelike, writhing luminescence.

It spreads through the narrow band of the now luminous atmosphere, up and down until a full half-circle is aglow. Suddenly, beyond the rim, there flames an immense wavering finger of fire, as though the globe were being consumed. A second and third finger of the sun's prominences shoot outward, until behind the earth they form a hundred fiery arches. With darkened glasses we watch as the sun's globular furnace, circled by the living halo of the corona, slowly emerges from behind the earth.

We know, in reality, it is we, and not the sun, that have moved.

The western limb of the earth has formed into a golden crescent. It is sunset on that part of the earth, for as we watch the area of the crescent slowly slips into the moon's paler glow.

The moon, slightly expanded, and half of her pitted surface pitilessly illuminated, floats ahead of us and to our right. We are not directed, we know, toward her present position, but to the place in the heavens where she will be at the journey's end — forty six hours later.

For the rest of the journey to the moon, we are told, we will have the sun with us on the earthward side of the ship — while beyond and all about us the impenetrable blackness of space will be broken only by the stars and moon. We will have night or day thc glory and power of light or the darkness of the depthless spaces — at the turn of the head.

But now the blazing heat of the sun is pouring with its full power into the observatory, and all but a few square inches of the ports must be shuttered.

With a multitude of new impressions beating on us, we return to our quarters. Our exhilaration is gone, the nervous tension relaxed. Tired by the swift succession of experiences in a crowded three hours, we are glad to return to our hammocks for the first night's sleep in space, twenty thousand miles above the earth.

CHAPTER X

IN SPACE

I

OUR first meal, after we awaken is truly an ordeal. Much food is spilled by the cook, still unskilled in handling the weightless containers; and as we sit round the improvised table in the centre of our living-quarters we make occasional abrupt movements that send our dishes flying swiftly across the ship. Fortunately they have been made unbreakable. We take a gleeful joy in testing the new qualities revealed by familiar things, and laugh uproariously at the awkward motions of our fellows who have been unable to master the new technique of dining.

After "breakfast," as we call it, the Chief calls a conference of the scientists. They have already made observations on the earth and the moon, and we learn that much valuable data has been obtained. The physicist has procured excellent motion pictures of the earth seen from twenty thousand miles as well as spectrographic plates of the cold stars of distant space. The astronomer who has seen them hints that they will create a sensation in astronomical circles when exhibited on the earth!

But now the scientists must decide on their programme for the exploration and study of the moon when we land. While they gather in the control room we search out the communications operator to learn the latest story from earth.

We learn that wild excitement prevails all over the globe. Cities have given way to unrestrained celebration, and the streets are filled with thousands of sky-gazers. The newspaper offices are constantly besieged for the latest scrap of information on our progress.

We cannot but feel now that this journey has served its purpose in the breaking down of racial jealousies. All nations have united in a communion of joy. All the earth's people share in the glory and the names of the men who form our party have already become internationalized — so intimately does the world know us.

Greedy for every minute permitted in the observatory, we leave the operator.

A shock awaits us. Our poor earth has shrunk greatly. From the great ball we saw at our last view it has become but an enlarged moon, some ten times the size of Luna as seen from the earth. We have been turning steadily away from our planet, and we see it now to the left and far behind.

The broad bosom of the South Pacific is in our line of vision, and we see more distinctly the moon-illuminated outlines of Australia and the white continent of the Antarctic. Slowly, with the calm dignity of a woman displaying her charms, the earth turns.

We are silent before this spectacle. We are suspended motionless in an eternity of space and time. The silence of the dark spaces is so overpowering that we dare not move. We hardly dare breathe. Except as indicated by movement of the earth, time has ceased. The darkness is no longer ominous, the stars no longer cold.

A great spiritual tranquillity fills us — a humbleness and a yearning for the continuance of this immense peace. Our being seems spread through the eternity that we can see. We realize now the full meaning of Einstein's "cosmic religion." Cities, empires, states; dreams and ambitions; conflict and confusion are infinitely remote, part of the dream-world of that slowly turning globe.

Abruptly the spell is broken as the scientists come into the room from their conference.

We besiege them with questions regarding the Lunar campaign. The Chief answers for them. Pointing off to the moon, he directs our attention to its illuminated north-west quarter, and to the Sea of Serenity. We are to land, if a landing is possible, on that great flat tableland. There, shaded from the sun for two days by a range of mountains, the Apennines, observations are to be made on the nature of the Lunar surface, the possible existence of a sub-surface atmosphere and of the prevailing temperature. The possibility of establishing an enclosed laboratory and a calling-station for a trip to Mars and Venus is also to be investigated as far as the short stay allows.

We alternate our gaze between the moon and earth. One world young, alive, teeming with plants and creatures of all orders; the other world lifeless, as far as we know — a burned-out cinder in space, perhaps a reminder to us of the fate of the earth.

II

Fourteen hours of our journey have passed when the Chief warns us to secure ourselves against a sudden shock. The ship is off its course, and the rockets must be set in operation again to return us to our predetermined path. In our quarters we grasp the hand-holds on the rear walls, and hear suddenly the now unfamiliar roar of the exhausts. The straps creak as we hold them tightly; we feel again the weight of acceleration press us to the walls. The pressure is the greater because we have gradually been accustoming ourselves to the experience of weightlessness. For a few moments we suffer an attack of vertigo. But the exhausts are again silenced, and after a few minutes we can weakly extricate our hands and feet from the straps.

The question of navigating the ship intrigues us, and we search out the Chief.

"Precisely," we ask, "how do you steer the ship where there is no air?"

" In principle," he answers, "the task is simple. Yet in practice, maneuvering in space is a delicate operation, and must be done with extreme care.

"If, for example, it were necessary to shift the course to the right, the rocket tubes on the left would be put into operation. This would throw the ship's nose to the right. Then, by using all rocket tubes, the course could be gradually inclined to the right. To move to the left the opposite would become necessary. To change the course in any of the other planes similar procedures would be adopted — rockets would be started on the side of the ship opposite to the desired change in direction."[1]

The same situation occurs in airless space. With no air or friction to retard the ship's forward motion the attempt to move to the right must first be executed by using the rockets on the left side, which would turn the nose to the right. The ship would then be moving sidewise without change of direction. If all the rocket tubes were now placed in operation a new impulse to the right would be obtained, and the course would be "forward right."

Hours later, as we wander between the peep-holes in the observatory, we note that our path is inclining toward the moon, and sense the gradual relinquishing of the earth's pull on the ship.

At the eighteenth hour of the journey we have covered 150,000 miles, and the earth is only six times as large as the moon as seen from earth. Behind and to the

1 The problem of changing the direction of a rocket-ship in airless space is really more complicated than is apparent. The best analogy may be drawn by the experience of a man on ice. Assume he is moving forward and wishes to turn to the right. To do this he twists his body, faces the right, and is soon moving in the desired direction. But be finds that in executing this maneuver he skids. In other words, he continues for an instant to move forwards. The explanation is that the ice has so little friction that the forward movement cannot be quickly broken. Let us assume the ice is to be absolutely frictionless. Then the attempt to move at once to the right would fail. Even though the man swung round and faced to the right he would continue moving in his original direction. If he now exerts his energy toward the right his motion will not be simply toward the right, but in a direction which is a resultant to his former motion and the new effort. He would move, so to speak, "forward right."

left of the earth the sun displays brazen charms so intense that we cannot view then for an instant with unshielded eyes.

But the moon, approaching swiftly, has doubled in size since the journey began. The scientist examines her scarred face greedily, and there is much excited conversation, and frequent exclamations that proclaim scientific discoveries. For now, as our path converges toward the moon's orbit, we are beginning to glimpse her other hemisphere — the half of our satellite that has been hidden to man since eternity.

As yet not much of it can be seen, for the moon approaching the full as viewed from the earth, has her hidden side in almost complete darkness ....

In six hours we will pass from the sphere of influence of the earth to that of the moon, and begin the 20,000 mile drop to the moon, on which we should land fourteen hours later.

On earth the western coast of America is slipping beyond our vision, and we catch our first view of Central Europe since the journey began. Directly behind lies the broken expanse of Asia and the moonlit sweep of the Indian Ocean.

With the passing of the hours and growing familiarity with the sensation of weightlessness life aboard the ship has assumed a fairly definite routine. We are too far away from the earth to receive its messages; and since we are in the glare of the sun, and cannot be seen from earth, all communication has ceased.

Of all the phases of life aboard the ship in "weightless" space, we have become accustomed least to the restrictions in our method of eating.

None of us have mastered completely the art of drinking by "squeezing" or of eating by "spearing"; and from the exclamations that emanate from the galley during the meal's preparation we gather that a new art of cooking, too, must be invented for the "weightless" spaces.

By the time we pass the neutral point of gravitation we have mastered the ability to walk without suffering undue bumps or bruises. But we look forward to the hour when the moon's gravitation will give us again a measurable feeling of substance.

III

Our electric chronometer registers 11•25 (twenty six hours after our start) when the first preparations for the landing are begun.

We have passed the neutral point of gravity with a speed of 6000 feet a second. If we permitted ourselves to be drawn unhindered to the moon we would crash upon her surface at more than two miles a second. If, however, with this speed, we attempted to manoeuvre to allow the moon to catch us in her gravitational net, and hold us as a satellite preparatory to landing, we should certainly fail, and shoot away from her.

We must, therefore, check our speed to less than a mile a second. If this can be achieved at a height of about 800 miles above the moon's surface we will circle her in four hours, and, checking the speed still further, slowly descend. At an altitude of 200 miles, an almost vertical landing on our satellite's bleak surface can be attempted.

Wondering just how all this is to be accomplished, we go to the observatory, and, securing ourselves with the hand-holds, watch the brilliant spectacle of a sunset on the earth, now behind and to the right of the ship.

Suddenly, as we watch, the heavens seem to shift to the left, and there comes again the roar of the rocket exhausts. So accustomed are we to the enveloping silence of space that the noise is as stunning as a physical blow. We see the earth, the sun, and the seats race across the skies until the nose of our ship points almost directly toward the earth. The ship has been turned end for end.

The roar becomes louder. At once the sensation of weight returns, although only a fraction of the amount we are accustomed to on earth. Descending carefully to the control room, we watch the crew at work making observations of the moon, which is now at our tail. The Chief is watching intently the dials that register speed and acceleration.

The commander is, nevertheless, not too busy to inform us that our rocket explosions are directed at a slight angle to the moon, or just ahead of the moon in its orbit. We will thus fight off the moon's growing attraction, and descend toward her at a gradually lessening speed, he explains, so that we can enter safely into an orbit about her.

Abruptly he turns to the call of the navigator, and we return to the observatory. The crucial moments of the journey are approaching. Two objects — our ship moving at one and a half miles a second, and the moon, moving in its orbit at approximately half a mile a second — must be maneuvered so that they will meet with almost no impact. Here is a gigantic task calling for the utmost skill and ingenuity of the Chief.

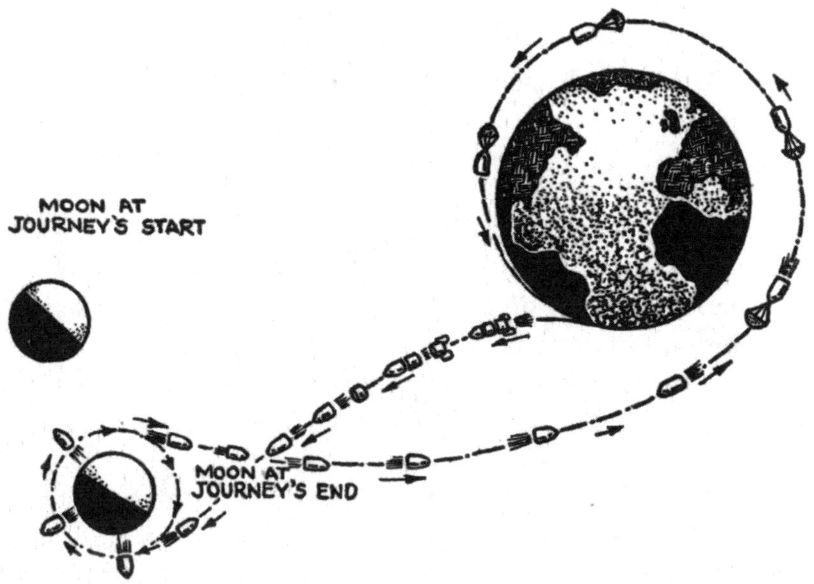

FIG. 4 THE PATH OF "TERRA" THROUGH THE HEAVENS
ON HER FLIGHT TO THE MOON AND THE RETURN TO EARTH

For simplicity the earth's position in its orbit is assumed as unchanged during the flight.

It will take only a little miscalculation to plunge us incontinently to destruction on the moon's lifeless surface; or, conversely, to miss meeting her entirely, shooting beyond her orbit. The latter eventuality would necessitate returning for another try. We would be obliged to change our course again and chase after the satellite, consuming meanwhile our precious fuel. After two or three such futile attempts at contact with the moon our reserves of fuel would be used up and the prospect of a safe return to earth endangered.

Peeping from the observatory port at the left, we see the moon just visible now "behind" us. Somewhat more than two-thirds of its visible surface is lighted by the sun. But these broken and tumbled masses of rock, the innumerable craters pock-marking its surface, inviting from afar, now take on a menacing air. We may soon be stranded upon that inhospitable world — devoid of water, air, or a single plant for food.

We are approaching the moon more and more slowly; yet we must cross her orbit ahead of her. Now only ten thousand miles away, her great plains and circular craters stand out in sharp relief, and the streaks of mountains throw their dense black shadows over miles of tumbled and jagged tableland.

The twin seas of Crises and Serenity are easily distinguishable, and in the centre of Serenity, where we are to land, we see the button-like crater of Archimedes, forty miles in diameter. To the left and below Archimedes the saw-tooth range of the Apennines, with peaks ten to twelve miles high, throws gigantic shadows into Serenity.

IV

The moon, as it approaches us, is now like a monstrous cannon-ball. How can our little car be directed toward a meeting with that gigantic bulk and avoid a disastrous collision? Closer and closer, comes its hurtling mass — filling the heavens. Then — after agonizing moments of doubt — we see her swing off to our right. Will we cross her orbit safely?

The danger is not yet over. We are abreast of her now — intercepting her path. Breathless hours pass as we shoot dizzily across her lifeless surface, only seven thousand miles away! In the control room we can hear the excited voices of the crew and the cool, incisive words of the Chief. Every second counts as the huge world rushes upon us. We are over her centre. She looms closer, swelling more and more. The sight of her dizzies us. The rockets are working furiously ....

Then, we understand, we will more than clear her. We are still going too fast . . . but now the moon swings through the skies to the left . . . we are beginning to circle at last . . .

We pass abruptly from her sunlit to her darkened side.

Now for the first time men are viewing the moon's hidden hemisphere!

But as we speed over the night-covered surface the view to our unscientific minds is disappointing. We see only gigantic black masses unrelieved by any lights that might give the faintest sign of life existing there. Another illusion persisting so long that even scientific authority could not destroy it, has been killed by our own observation — the moon is lifeless.

Quickly we pass into the moon's shadow. Above her black surface tremendous tongues of flame, green and vividly red, lick hungrily into the skies. We are viewing an eclipse of the sun with the beautiful corona displaying to the full her dazzling glories. And there — off toward our left, still above the moon's surface — is a thin crescent of silver — the earth!

But we move on; the earth is lost to view though the corona remains behind us to blind and dazzle with its suggestion of hidden paradises. We are lost in a bewildering display of colour and power and glory. Though we have been here in the observatory for hours so many exciting events have occurred that time has forsaken its meaning. Children of the cosmos, we have been treated at last to the picture of nature's magnificence.

The black Lunar surface passes slowly beneath as we move gradually spiralling towards it. In an hour we will again emerge into the lighted side.

We descend to our living-quarters after our new experiences. Since we are sheltered from the earth by the moon we will be out of view of our homeland until we are again on the moon's sunny side.

Lost as we are to view, great anxiety must prevail on earth during these hours ....

We snatch an hour of sleep, to prepare for the exciting moments of the landing.

CHAPTER XI

THE RETURN TO EARTH

I

When we awaken we become immediately aware of the milky glow at the porthole and the softly diffused sunlight through the room. We are again on the moon's sunny side. A landing will be made in five hours.

We are now about 1500 miles above the moon's surface, and our ship is moving about her sidewise. Our tail is toward her surface, and the nose and observatory are pointed into space.

We hasten to the observatory — noting that a semblance of gravitation is now exerted on us. We still wear iron shoes, and we can walk erect, with the dignity of explorers of the great spaces.

The moon is just visible behind us from the observatory port, but to the left is the crescent of the earth, the broad bosom of the Pacific radiant in the sun's glow.

When we descend to fortify ourselves with food for the last hours of our stay in space we learn that our space suits are to be tested immediately after the meal, so that an exploration of the Lunar surface can be made when we land.

Here we have a new source of fun — trying on these great ungainly suits of metal — encasing ourselves in them; adjusting the oxygen tanks on our backs and pulling faces at each other through the glass windows at the eyes.

We resemble some fantastic pictures of deep-sea divers, or gallant knights of old, ready to explore the moon for romance and adventure with possible Selenites!

The success of the space suits demonstrated, we return to the observatory to watch the earth and sun swing about us. Through a peep-hole our physicist grinds away enthusiastically with his camera, catching all possible views of the earth.

Regretfully we must miss the sight of the landing, for all the ports that give views of the Lunar surface are used by our crew, intent on the last delicate maneuvers.

We are circling the moon at 3500 miles as hour, and have again entered her darkened side. Our speed must be swiftly decelerated, so that when we again emerge into the moon's lighted half we can drop gently to her surface. A complicated set of maneuvers are necessary, and disaster may yet result.

With this knowledge, and despite the utmost confidence in the Chief, an air of anxiety pervades the group in the observatory.

We watch the skies impatiently, as though they could give some hint of the success of our venture. But the sun and earth swing away from us to disappear on our left, and we face again the cold star-dusted skies . . . Slowly, as time passes and as stars stream across our view, we watch anxiously for the first glimpse of the sun's corona on our right. We catch it finally — the brilliance grows until our astronomer informs us that we should again be directly over the Sea of Serenity.

But there is no sign of a landing. The full globe of the sun appears in our line of vision, moves off to the left, and gives way to the earth.

We become alarmed. Has something gone wrong? Cannot the landing be made? We wait fearfully, not daring to disturb the Chief at his terrible task. Slowly the earth floats across the sky, and we watch the mountain chains of Asia pass from day into night; and then we are again pointing into space.

Four hours have passed, when the Chief, grim-faced, appears in the observatory. He looks exhausted from the unremitting labour of the past hours.

"I'm afraid there will be no landing, gentlemen," he says slowly. "This maneuver takes more fuel than we have allowed for: and the ship will not stand the jar of the impact. I should like to meet you gentlemen in conference below."

Gravely the men file down to the control room, while we wait anxiously for the verdict. Is disaster in store for us?

When the Chief appears, half an hour later, he tells us that the ship will be turned about so that we are once more parallel to the moon's surface. We will circle it again to enable the scientists to make their final observations, and then start immediately for the earth.

Hardly expecting this, the news electrifies us. We realize at once that empty space has enveloped us so that we had come to think of the earth as a remote globe we had long ago known, but to which we would never return.

A wave of homesickness strikes us, and again comes the disturbing melancholy and a tendency to tears. We must control ourselves.

We bombard the Chief with questions, and despite his weariness he answers them patiently. The lack of any atmosphere on the moon makes a gradual landing extremely difficult. The original plan was to drop down tail first, sustaining ourselves by rocket exhausts. When we were close to the surface we would maneuver so that the axis of the ship would be parallel to the moon and a long glide could be made for the last few hundred feet. But the power used for the maneuvers has already eaten so deeply into our reserves that a landing cannot be chanced.

The rewards of the trip have already been great enough, the Chief says, smiling. We shall return to earth, and by the stimulation of our success and the knowledge we offer work can be hastened on a ship equipped for a safe landing.

Twelve hours have passed. We have circled the moon, and at last we are leaving it behind us. Our course is directed into the heavens to a place at the right of the earth — where in forty-eight hours we shall again meet our great Mother.

II

Now that we are returning we begin to long for cities, green fields, and flowing streams. We have met harsh Nature, and seen her in a pitiless as well as a splendid garb. But we cannot endure her nakedness for long. She is too splendid and too cruel. Her tones are too sharp. We want softness again — grass beneath our feet, softened sunlight, and gentle winds.

Restless on leaving the earth, we are impatient to return; and as the moon shrinks in the skies and the earth grows swiftly homesickness grips us. We are physically tired, and enjoy the first real sleep since the journey began . . .

When we awaken we have passed the neutral point of gravitation 20,000 miles from the moon, and have entered the earth's influence. The rockets had been started to enable us to free ourselves of the moon. Now they are silenced again, and no power will be needed until the beginning of landing maneuvers at the earth.

The earth is assuming a pleasing bulk, and the growing sharpness of detail of the continents and oceans suggests to us pleasures whose value we had not known until this moment.

At dinner the Chief announces a surprise. To compensate for the failure to explore the moon we can, if we wish, explore space itself! We stiffen with interest.

Immediately after the meal, he says, we can "walk into space," two men at a time, in our space suits, and enjoy the unique experience of standing in a vacuum with the nearest ground 25,000 miles away.

The men are excited. We hurry through the meal, and in our room put on the space suits — which had been regretfully packed away.

Encased in metal, the first pair to go are given metal cables with which to pull themselves back to the ship.

They are ready now. With thumping hearts we watch as the Chief sends a supply of air into the lock which leads from our room.

Can it be that these men may stand suspended in space? We are travelling, we understand, at a mile and a half a second. Won't they be left behind? And can those thin metal wires hold them securely to the ship moving at this great speed?

The air-lock is ready. The inner door is slowly opened, and the two metal men step into the lock, the door closing behind them. We hasten to the observatory to watch this miracle. We see the men already in space. They stand upright, somewhat awkwardly, holding tightly to their wires. Then one of them releases it! Is he lost? The wire remains floating in the void, and beside it floats the man travelling with us at four thousand miles an hour.

They seem to enjoy pulling themselves about on the rope and then letting go of it suddenly. Masters of space, they wish to use their power to the utmost.

Finally they disappear into the ship, and two more emerge to test the wonders of weightlessness. Although there seems to be no danger, the thumping in our hearts remains, and becomes increasingly violent as we return to the room for our turn. We would like to decline — but with the newspapers of the world expecting an "eye-witness" account of a stroll in space we must go through with it.

Nervously we encase ourselves in space suits and take our places in the air-lock. Gradually, as the air is exhausted, the flexible metal of our suits begins to creak and stiffen. Finally, at a signal from the Chief, we open the door to space.

Before us stretches the great empty gulf like a bottomless well. We step back in terror! But you have passed by me and nonchalantly step into emptiness. I expect to see you hurtle out of sight, followed by unheard despairing cries. But, grasping your rope, you stand calmly outside and beckon to me.

A step, then another, and, closing eyes to the dizzy emptiness, I force myself into the open. But the miracle has occurred! I do not drop. I feel nothing. As though I were afloat on water, I rest on the wings of space.

Beneath us the blackness, shattered by cold stars, extends to infinity. Behind lies the bleak full face of the moon, while in front and to the left the golden crescent of our earth beckons.

We move an arm, then a leg. The motions are without sensation of any kind. It is as though we have no limbs, but are disembodied spirits of the great spaces. Only the pressure of metal hands on the wires assures us that we are physically alive.

You release your wire, and it remains beside you as though held by an invisible hand. Gaining courage, I release mine. There! I am cut off from the ship entirely — I am adrift. If anything should happen now the ship would go on without me, and I should remain here eternally, a wanderer of apace. The idea does not appeal so I regain my wire and slowly pull myself back to the ship.

Later, when we can question the Chief, we ask him to explain this apparent paradox. Why did we remain thus in space, beside a swiftly moving ship, and neither fall nor be left behind!

Very simple, the Chief explains. There is no friction in space, and in this area no appreciable gravitation from any heavenly body. The only force acting on us is the velocity of the ship, of which we have partaken. When we step into space we still retain that velocity, and we float along with the ship as though we were still part of it.

Anticipating our next question, he illustrates the difference between this and jumping from a moving aeroplane.

The aeroplane, to maintain its speed, must constantly be supplied with power to overcome the resistance of the air. When we step from the aeroplane, therefore, we have only the aeroplane's velocity, which is rapidly diminished by air friction. Further, the earth pulls us to her. So, describing a parabolic course, we would fall rapidly to earth. But moving, as we are, without power, there is neither the air friction nor sufficient gravitation to separate us from our ship.

III

For the remainder of the trip we settle down to the preparation of our reports. We learn that communication with the earth has been already established, and the news of our return has completely upset the ordered life of the globe.

Pressed by the earth's people to give an indication of the place of landing, the Chief has mentioned the Bay of Biscay, and he has asked the Commission to have a number of fast cruisers scouting the Bay to pick us up when we land.

As we fall into the earth's gravitation our floor is again shifted to the side-wall. We are being pulled by the earth in the direction of the ship's travel, and to hold our present floor would mean literally falling on our heads.

Seasoned astronauts, we make this change easily and watch eagerly the earth's approach. We are only five thousand miles away when our ship turns abruptly end for end, tail first, and we begin hurtling toward the earth.

Swiftly it rises toward us. With rockets firing furiously at the tail to slow our speed we are plunged toward her surface. Like a magnet, at an altitude of five hundred miles, we are pulled into an orbit about her.

Swiftly we circle the globe, as continent, lake, river, and ocean speed beneath us. We pass swiftly from night into day, then into night again as the earth flies under us. Now we have broken our orbit, and, checking our speed again by the rockets, we speed tail first at an angle toward the earth.

The great parachute at the nose is released, and is ready to grip the air and retard our speed so that a gentle glide, sustained by our wings, will bring us to the earth's surface.

With a great roar we feel the first impact of the rarefied atmosphere. The parachute fills out. There is a sudden jar. Incontinently we are thrown about.

The ship is already warmed by the friction of the air. For hours now we have been sensing increasingly the gravitational pull of the earth and our decelerated speed. Three of the men are ill. The unaccustomed sense of weight has been too great, and they lie on their cots, pale and hardly breathing.

On the insistence of the Chief we hurriedly strap ourselves into our hammocks.

We envision the approaching earth. It cannot be more than a hundred and fifty miles away. Possibly cities can already be distinguished. Certainly we have been seen from the earth — the crowds are waiting — the coast will be lined with ships of all kinds loaded with people . . . .

We think of the beginning of the trip . . . or perhaps this is the beginning. As we close our eyes we are not certain, for a sudden weakness grips us . . . we can hardly breathe . . . . Our heads are filled with madly pulsing blood . . . . We hear again the roar of a thousand cannons . . . .

There is a pounding in our ears. Suddenly we smell the salt of the sea. We lie with eyes closed, and then weakly open them, to look through the open air-lock to the Atlantic, on which we are gently rolling.

Above us stands the Chief. His face is pale, but he smiles

"The journey is over!"

PART III

NEW WORLDS

CHAPTER XII

BARRIERS

I

Our journey is over; we must return to earth and its realities.

A picture of a space flight was presented in the previous Part that assumed the surmounting of all difficulties in the construction of a space vehicle and ignored the many dangers that will beset daring astronauts. Now we must face them — the barriers to our immediate conquest of space and see if any are unconquerable. Only then will the reader acquire an understanding of the outlook for the space flight, and in addition have full appreciation of the enormous task that astronauts have set themselves.

The successful space flight demands that men and materials must perform feats of strength and endurance never before asked of them. This has already been suggested in the preceding chapters. The idea will be developed more fully in this discussion in the hope that the conquering of the barriers, through the years, will meet an intelligent evaluation of the accomplishment.

II

The space-ship Terra was assumed to be propelled by fuel powerful enough to lift it beyond the earth's attraction and to enable it to circle the moon and return.

For the verisimilitude necessary to fiction the landing on the moon was pictured as being impossible at the time, the point being that new forces may develop to restrict the expected achievement, despite all the care with which a journey into the unknown is planned.

To return to our fuels, the hydrogen and oxygen, and oxygen and alcohol used to propel Terra, although the most powerful fuels known, yield in reality energy barely sufficient to permit a space-ship to make the interplanetary journey. Until a new fuel is developed and its action controlled, or our technology of construction improved, the space flight will remain an exciting but impracticable goal.

In simple terms, to lift a one-pound weight beyond the earth's attraction requires the expenditure of about 21,000,000 foot-pounds of energy. If the force necessary to overcome air resistance within the atmosphere, to check the ship's speed so that it my safely approach the moon, and then to return to earth is included, a pound weight will need approximately 30,000,000 foot-pounds of energy.

The mixture of one pound of liquid hydrogen and oxygen yields on perfect combustion some 5,500,000 foot-pounds of energy. If the fuel were all exploded at the earth's surface six pounds of fuel could send one pound of space-ship on its Lunar journey; and only the fuel for steering and for the return to earth need be carried. Theoretically this procedure would solve the fuel problem.

But this is merely a restatement of the Jules Verne projectile. It would be equivalent to shooting the rocket from a gigantic gun, so that when it emerged it would already have the required seven mile a second speed. And since this would result in the complete extinction of the occupants, either by roasting or crushing, it must be abandoned. The space-ship must be accelerated gradually, and to do this the fuel must be carried and used little by little.

Since the fuel must also lift itself part of the way into space, and can only use a small part of its energy for lifting the space-ship, at least 250 pounds of fuel must be carried at the start for each pound of space ship or payload that is to reach the destination. The pound of payload allowed must include not only the passengers, equipment, and supplies, but also the weight of fuel chambers, auxiliaries, and the ship's hull.

The 250 to 1 ratio obviously makes the problem incapable of solution. Our present knowledge of metallurgy does not provide materials of which to construct the ship that will be light enough, yet strong enough, to fit these requirements.

Unless we can develop, on this basis, a fuel providing immensely more energy per pound than the oxygen-hydrogen mixture the construction of a space craft cannot be thought of. This obstacle was recognized quite early by rocket experimenters. The step rocket was their answer, offering a solution of the difficulty.

III

Only when we return to the step rocket do we perceive the possibility of a present conquering of the fuel barrier. The step-rocket principle is an ingenious but simple device, and is illustrative of the ability of our astronauts to overcome their difficulties.

Terra was built on the step principle — equipped with two auxiliary rockets, the first giving to the ship a speed of three and a half miles a second and the second after the empty hull of the first had been thrown off increasing the speed to seven miles a second, whereupon it was detached.

As illustrative of the step-rocket principle, the space-ship proposed by Hermann Oberth may be used.

For an ascension into the outer spaces, and the attaining of freedom from the earth, Oberth suggests a three-step rocket. Each step would raise the speed of the ship two and a half miles per second, thereby giving to the pay load a final speed of more than seven miles per second.

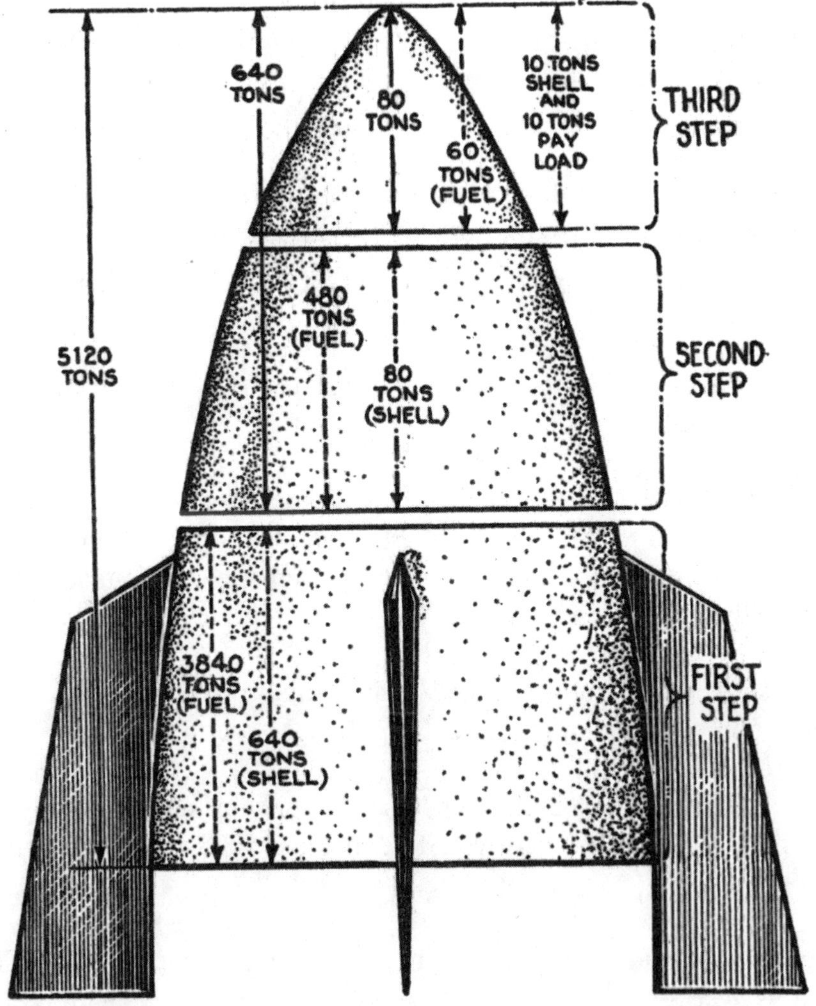

FIG. 5 DIAGRAM TO SHOW THE DISTRIBUTION OF WEIGHT OF A THREE-STEP ROCKET PROPOSED BY HERMANN OBERTH

Oberth calculated that the only feasible ratio of fuel to pay load could be 3 to 1, instead of the 250 to 1 ratio made necessary by our original figures. He therefore designed a rocket-ship to weigh at the start 5120 tons. The first and largest step, as shown in the illustration (Fig. 5), would have 3840 tons of fuel. Its total load, 1280 tons,[1] would consist of 640 tons representing the weight of the shell, operating equipment, and rocket motors for this step, and 640 tons consisting of the second and third steps. Based on this 3 to 1 ratio, a construction that is possible, the first step would give to the entire ship a speed of two and a half miles a second, the 3840 tons of fuel being meanwhile consumed. The empty 640-ton shell would then be detached.

The ship now weighs only 640 tons. Of this 480 tons is fuel for the second step, the load, 160 tons, consisting of 80 tons of shell and motors for this step, and 80 tons representing the weight of the third step. The burning of the 480 tons of fuel will increase the speed of the ship to five miles a second, whereupon the hull is detached.

An 80-ton ship will remain. It consists of 60 tons of fuel and 20 tons of load, of which 10 tons are hull and 10 tons are the passengers, supplies, and equipment. By burning the 60 tons of fuel the speed of the ship will be increased to seven miles per second, freeing it from the earth.

By the clever device of the step rocket an interplanetary journey is thus made an engineering possibility.

Nature, however, has not been cheated. It will be observed that a total of 4380 tons of fuel was used to free 20 tons of the earth's attraction — a ratio of more than 200 to 1. What was done, merely, was to design the ship so that the ratio of fuel to load always remained 3 to 2.

The ship described could only be freed of the earth's attraction. It has no fuel to circle the moon or to return to earth. To make the space flight really possible, therefore, one more step must be added. Preserving the 3 to 2 ratio, the new step, which would now become the first step, would then weigh approximately 36,000 tons, of which about 30,000 tons would be fuel and 5000 tons the weight of its hull.

In summary it is evident that a ship of 40,000 tons, almost the size of the Leviathan, is necessary to bring a 10-ton pay load to the moon and return. Only four passengers could possibly be accommodated on this huge vessel.

1 1280 tons is one-third of 3840 tons.

With present fuels, therefore, the interplanetary voyage is an undertaking requiring a huge fortune. The cost of the ship, including the preliminary experiments and tests, may well reach $100,000,000 (£20,000,000).

These dry statistics of energies available and energies required presume, furthermore, an almost perfect use of the fuel. There is no margin left for chance, for human errors, for unseen contingencies. The return to earth, even on the basis outlined, would probably be carried out by desperate explorers whose reserves of fuel had been almost exhausted by unforeseen contingencies, and who realized that only superhuman skill in the delicate maneuvers of the landing would bring them safely to the earth's surface.

For a journey to Mars or Venus, or a landing on the moon, the requirements in fuel would increase the initial weight of the space-ship to a truly tremendous total. Half a million tons would not be an exaggerated estimate of the weight of a ship that is to descend to the Lunar surface, while for a voyage to our sister planets the requirements may well be double that figure.

Despite these staggering figures the problem is no more incapable of solution than were the desperate efforts of Orville and Wilbur Wright to find a machine to give them a powered flight in the air for a few hundred feet. Only perennial doubters will assert, with the art of space navigation as young as it is, that we face insuperable difficulties.

Even though our present fuels do not make the complete interplanetary itinerary possible, they do make possible a flight into outer space, several thousand miles above the earth, and a safe return.

Such flights must of necessity precede the more ambitious moon journey, for they will acquaint our astronauts with the actual conditions in space, the technique of interplanetary navigation and the methods of making the landing at the journey's end.

Such flights will be more ambitious and more daring as our practice in navigation in outer space grows and our knowledge of explosive fuels increases. Scientific men have been thinking seriously of harnessing known compounds for use as rocket fuels for less than ten years. With the ever-increasing interest in rocket experimentation and modern progress in fuel technology great strides are inevitable in conquering the barrier now presented by our incomplete knowledge of the best fuel for the spaceship.

As will be pointed out more fully in a later chapter, we have, in fact, only begun to survey the field of fuel possibilities. In the energy of radium, in the infinite radiant energy of the sun, and in atomic energy there reside reservoirs of power which by comparison make all of our present fuels seem impotent. It is by no means beyond possibility that the discovery of a way to utilize one of these will solve the problem of fuel for space flights, even though a new and more powerful explosive compound is not devised.

IV

In the voyage of Terra we assumed that our crew were concerned only with the navigation of the ship in empty space . . . . Actually they will have other serious concerns and dangers. Among them are the meteors.

As "shooting stars," meteors often delight us. They provide occasionally, on clear nights, a brilliant heavenly display as they shoot recklessly across the sky. But to the prospective navigator of space they present a menace, and, ironically, the smaller they are the more dangerous.

Most meteors are lumps of iron, with possibly a mixture of nickel or cobalt, and are presumed to be remains of exploded comets or particles set free in the upheavals of other solar systems. They vary in size from specks of dust to great blocks weighing fifty tons or more.

For the most part they follow definite orbits about the sun, as do comets and the planetary bodies. The two major meteoric groups — the Leonids and Andromedids — have orbits so well known to astronomers that their arrival near the earth can be predicted to a day.

Travelling in swarms of hundreds of millions, they swing from the limits of the solar system toward the sun, circling it and returning to the extreme limits of their orbits. Their path crosses the earth's on each journey, and when the earth intercepts them, notably in August and November, we have brief dazzling displays of shooting stars.

If the meteors were all in definite groups their coming could be predicted with certainty, and an interplanetary space expedition need only be planned for a period when the space lanes were free. But unfortunately many of the meteors do not travel in swarms, and others do not follow the usual orbits. Even the earth is said to be bombarded regularly with ten million of these errant meteors each day.

Space, although a vacuum, is thus filled with hundreds of millions of these metal and stone particles, large and small, travelling at speeds ranging from fifteen to forty-five miles per second! A meteor the size of a pin's head could crash through the best armor — plate of a space-ship admitting to the ship's interior the terrific cold of space, and providing an escape for the precious oxygen. With no means of protection the space-ship might well be riddled by small meteoric particles, even if a larger one did not strike it with such stunning force as to wreck it.

The danger presented by meteors does not lie in their number or size, for although ten million do strike the earth each day their actual density in space is quite small.

Dr. Goddard, by assuming meteors to be 250 miles apart, has estimated that on a journey to the moon a ship one foot in diameter would have only one chance in a hundred million of being struck.[1]

A fair-sized space-ship would have a million to one chance of escaping without a meteoric collision.

The meteor question is a source of great fear and acute mental uncertainty rather than immediate danger. Although the million to one odds may in theory serve to assure the crew of a space-ship of their safety, they will have in actuality many terrible hours watching particles flying uncomfortably close. And they will know that there is no possibility of maneuvering in space quickly enough to escape should a meteor bear down on them.

Even if an oncoming meteor could be sighted ten miles away, its speed of thirty to forty miles a second would permit only a fraction of a second to remove the bulk of the spaceship from its path.

The danger from meteors must be chanced by our explorers. It is altogether possible that further observations on meteoric swarms may prove that a journey made during certain months will find the heavens comparatively free. Or rockets exploring space may be able to chart the paths of meteors closely enough to indicate definite channels through the heavens that are safe enough for all practical proposes.

V

The great interplanetary spaces are filled not only with dashing meteors, but also with an unseen force that may be immensely more destructive and less easily

1 Dr. Robert H. Goddard A Method of reaching Extreme Altitude, Smithsonian Miscellaneous Collection, Vol. lxxi No. 2 (1919).

avoided. Shot out from giant aggregations of solar bodies there are extremely penetrating radiations of energy that reach our solar system and threaten the extinction of all unprotected living matter. These so-called cosmic rays, created in the birth throes of distant worlds, fly with the speed of light through all space, and are so penetrating that eight feet of solid lead is necessary to block them.

On the earth's surface we are protected from these rays by our 250-mile blanket of air, but in the vacuum of space we will be exposed to their full intensity, the effect of which we do not know.

Because they are a radiant energy of such an extremely low wave-length, ten thousand times shorter than the X-ray, and because we have no means on earth to produce them, we have never been able to approximate or to measure the cosmic ray's power. Present theories propound that even the protons and electrons, the primal building-bricks of matter, are recklessly created and destroyed in the unspeakable furnaces that give birth to the cosmic rays.

By actual experiment cosmic rays are found to exert their tremendous energy in the destruction of matter. Dr. Robert A. Millikan, who discovered their presence, found that even at sea-level the rays are able to break up 1.4 atoms in every cubic centimeter of air each second, and that they must break up millions of atoms in our bodies. Their effect is similar to that of radium, which destroys animal tissue by means of its extremely short radiation. But the gamma rays of radium are only one-hundredth as powerful as the cosmic rays.

Such is their effect on us even when we are shielded by our blanket of terrestrial atmosphere.

But Millikan found by the records of measuring instruments that the intensity of the cosmic rays increased rapidly as one ascended into the thinning atmosphere. The deductions from Millikan's conclusion were, therefore, that in outer space, without atmospheric protection, one would be exposed to their full and complete power. It is possible that all living things in free space would be destroyed unless encased behind walls or solid metal several yards thick.

With the menace thus pictured at its worst it is now possible to search out the actual danger. In the first place the exact physiological effects of the full power of the cosmic rays are not really known. No reputable scientist will predict with absolute certainty that they will entirely destroy animal matter, or that they cannot be reasonably deflected.

When Professor Auguste Piccard ascended ten miles above the earth in a balloon on May 27,1931, neither he nor his assistant suffered any noticeable effects from cosmic rays. At the altitude they reached the air density is but one-tenth of that at the sea-level, and the intensity of the cosmic rays should have been proportionately increased. Professor Piccard's experience offers evidence, therefore, that the cosmic ray danger may be much exaggerated, and that the natural protection of a double-walled space-ship may be sufficient.

The unmanned meteorological rockets will surely discover the maximum intensity of cosmic rays in outer space; and to determine their effect upon living matter a rabbit may be sent aloft to be subjected to their full power.[1]

The barrier presented by the cosmic rays is yet to be determined. If they prove to be as destructive as their physical qualities would indicate the programme for the space flight will be faced with a serious obstacle. On the other hand, it is possible that they are no menace, and because of their extreme velocity and great penetrating power the cosmic rays may actually be put to work freeing the titanic energy contained in matter — and so provide space explorers with a supply of unlimited power.

CHAPTER XIII

BARRIERS continued

I

We have pictured our space travellers as yielding only temporarily to the terrors of weightlessness; on the cruise of Terra they recover quickly enough to enjoy the exhilaration that the freedom from heaviness brings.

In actuality it is to be doubted whether the pioneers into space will escape so easily. The adjustment of our bodies to the sensation of weight over millions of years cannot be altered in an hour. And although the exact effects of the absence of weight will not be known until it is experienced, conjectures on the subject may profitably be attempted.

Of the two effects — mental and physical — the former; as the more indefinite and the less easily settled, will be considered first.

The passengers of Terra, leaving the earth, found that as they were accelerated

1 It has already been announced that Dr. Darwin Lyon, an American physicist, intends to send a rocket carrying mice and birds high into the atmosphere to test the effect of cosmic rays upon them.

into space their weight had increased four-fold — our 150-pound man supported an extra load on his body of 450 pounds. A few minutes later, when the rocket's power had been shut off and the ship was virtually falling through space, the passengers were entirely weightless. These changes of sensation are quite violent, and their effects cannot be passed over lightly. Many of the sensual impressions that govern our nervous — and therefore mental — reactions are based on the familiar feeling of weight — which means heaviness, or substance.

To shatter rudely these habitual sensations is certain to create in the mind great mental disturbance accompanied by a feeling of considerable apprehension.

The sensation of falling through space[1] is bound to be very real to our weightless explorers. Unalleviated, this feeling may well create such a state of mental anxiety, especially after the violently disturbing experience of the start, that a permanent state of mental tension might result.

It is not alone the feeling of being totally unsupported, which the sight of the floors and walls cannot banish, that will cause such mental hardships, but also the knowledge that the ship, without any indication of motion, is tens of thousands of miles from "land." The sensation of falling freely may well become associated in the mind with the falling of the ship, and the explorer may conceive with terror that he is experiencing a sheer drop through practically unlimited space.

If this mental tension is a result of uncertainty with relation to his position in the ship, it can be relieved in part by the magnetized steel shoes worn by the passengers on Terra. A sense of contact with the floor will always be present then, and, although there will be no feeling of weight, the terror of being unsupported will no longer exist to the same degree.

An ingenious suggestion from the German interplanetary group offers one approach to a solution of the entire problem of weightlessness. They propose no less than the creation of a state of artificial gravity by centrifugal force. The living-quarters of the ship would be in the shape of a cylinder with the axis coincident with the axis of the ship. The cylinder would be revolved. The outer periphery becoming the floor, there would be produced a centrifugal force on objects on the floor that would create the sensation of weight. In other words, the revolving floor would pull objects to it just as our earth, by gravitation, attracts and holds everything at its surface.

1 A common experience in dreams, perhaps harking back to the falls our arboreal ancestors experienced.

Our space travellers and their furniture would thus be distributed round the walls of the revolving tunnel like room, their feet outward, their heads pointing radially inward. If an appropriate speed of revolution is maintained the exact duplication of earthly weight would result.

The difficulty is two-fold. The necessity of heavy machinery and power required to operate this "weight-producer" continually would hinder a solution of the vexing problem of fuel; and the size of the chamber necessary to produce a comfortable rate of revolution would require a ship of enormous dimensions. If the chamber is to revolve as infrequently as seven and a half times per minute it must have a diameter of a hundred feet! And if the size of the chamber were reduced the speed of revolution would have to be proportionately increased, causing acute dizziness, and other bad physiological results.

The idea of such a great chamber, most of which is empty, built for the sole purpose of creating artificial gravity, cannot be entertained until the fuel problem is settled to satisfaction. Meanwhile, we must regretfully abandon this pleasant device and be content with the aid afforded by magnetized shoes.

II

A suggestion made by a physicist, Mr. Philip Barr, also merits consideration.[1] Mr. Barr proposes for the dual purpose of relieving the terrible weight caused by the initial acceleration and the correspondingly bad effects of weightlessness that the ship be accelerated slowly but uniformly from the earth during the greater part of the journey, producing a sensation of weight comfortable at all times. The floor of the cabins would be at right angles to the direction of flight

Instead of a great initial acceleration to reach the seven-mile-a-second speed as quickly as possible, Mr. Barr would increase the speed uniformly twenty two miles per hour each second. At the end of the first second of travel the speed would be twenty-two miles per hour; a second later it would be forty-four miles per hour; at the end of ten seconds 220 miles per hour. Since this acceleration is equal to that of gravity at the earth's surface,[2] there would be at the start the feeling of doubled weight instead of a four-fold increase. But as the earth's attraction gradually diminishes the weight sensation would approach that we are normally accustomed to on earth.

This suggestion promises to eliminate at once the two harmful concomitants of the flight. But Mr. Barr himself willingly admits its limitations. In the first place,

1 Philip Barr, "New Worlds to Conquer," Outlook October 8, 1930
2 Thirty-two feet per second per second.

such a slow acceleration would be quite wasteful of energy, for the fuel would be working at a low rate of efficiency during a great part of the journey. At the end of one day, furthermore, the speed of the ship, with this acceleration constantly maintained, would reach 500 miles per second!

Hence, in order to make a landing on any planet, this terrific speed would have to be retarded by rocket recoils and reduced to something like one to two miles per second.

Mr. Barr admittedly bases his plan on a fuel of power so great that no thought need be given to its expenditure. Because we have no such fuel we must turn regretfully away from this solution. We see rise at every point the limitations of our present fuels, to balk our efforts to make the journey equal in comfort to passage in a liner.

Assuming for the time being that weightlessness cannot be avoided, and that serious mental disturbances may afflict our astronauts, we may now turn our attention to the physiological effects.

These are more easily determined, for we are already conversant enough with the body functions to single out any that are dependent on the action of gravity. We know what organs must hang in a definite position and the muscular actions that are associated with the expenditure of energy in moving about.

According to the authorities investigated, practically all of the physical activities of the body are dependent on muscular and osmotic actions, rather than on gravity. The processes of breathing, assimilation and digestion of food, the throwing off of waste products, and the circulation of the blood would all function even were there no weight. Captain Hermann Noordung, the German engineer, state as a result of his investigation that "all the processes important to life prove to be wholly independent of the position of the body, and are performed equally well in an erect or lying position . . . ."[1] The only physical danger Noordung perceives, and he is supported in his view by Hermann Oberth, is that atrophy of the muscles is likely to result from disuse. The passengers of the space-ship would need little energy to perform any physical act, and a too-careful conservation of muscular force might so soften the muscles as to make them non-responsive when the space travellers returned to earth. "However," Noordung adds:

1 Captain Hermann Noordung, The Problems of Space-flight.

> it is probable that by systematic exercise this could be successfully provided for .... Presumably the only organ influenced by the absence of weight is that of equilibrium in the inner ear. Even this would no longer be needed, for the idea of equilibrium ceases to exist in the absence of weight.
>
> In every position of the body we then have the same feeling; "up" and "down" lose their usual significance, with regard to the surroundings; the floor ceiling and walls of a room may each be considered as merely a wall.[1]

Noordung would, with justice, have travellers in space walking from floor to the walls and ceiling with the careless abandon of a wandering fly. With magnetized shoes this would indeed be possible — in the state of weightlessness all directions would completely lose their meaning.

We leave, then, this rather fascinating question of the effect of weightlessness with the realization that the mental effects must be chanced, while the physical ills are likely to be small providing the body is kept in an otherwise healthy state.[2]

III

We have still to consider several more of the barriers in the way of space travel. Of these, the circumstances that may prevent man's existence on other worlds, and thereby bar him from reaping the rewards of his hazardous journey, will be considered in a later chapter. The question of his life and comfort in space and the charting of his pathway through the heavens merit immediate attention.

The primary requisites for the continuance of life — food, drink, air, a pressure to equalize that to which the body is accustomed, and a moderate temperature — must, as we have seen, be supplied to our explorers during the journey. This is true not only for the occupancy of the sealed tightly spaceship, but also while it rests upon the surface of a strange and unknown world.

Because of the limitations of space and weight the essentials for life and comfort must be as compact in form as possible, and restricted in quantity. Common sense dictates a supply of necessities to last several times as long as the journey will probably take, to provide for unforeseen contingencies, and they should be doled out in not over-generous quantities. The space flight, like all explorations, is only for lean, strong, and abstemious men.

1 Op. Cit.

2 It has been suggested to the writer that the removal of gravitational attraction might cause the heart to beat very much faster, perhaps fatally so.

Food will probably come in concentrated packets — in a form that offers the essentials for good health with the least amount of waste matter. Food will mean digestible bodily fuel, with the requisite amount of rubble, the proper number of calories, and a generous assortment of vitamins to ward off the onslaughts of scurvy and malnutrition.

Cubes of bouillon, dried meat and fish, concentrations of vitamins, as in orange and tomato juices, and evaporated or powdered milk — will probably be the diet. Quantities of water would also have to be carried.

For his Antarctic expedition of 1929-30 Admiral Byrd obtained the assistance of a great number of dietetic experts to plan the expedition's food-supply. Science will be similarly called upon to create new synthetic foods for an interplanetary journey. Our hypothetical International Commission will probably consider the question of food-supply as requiring its earliest attention.

The science of artificial ventilation has already advanced far enough to ensure complete success in adapting it to a space-ship. Oxygen would be drawn from the general fuel-supply and used over and over, while water vapor, nitrogen, and other gases necessary to provide the proper pressure within the ship would be carried separately in liquid form.[1] The air stream after passing through the living-quarters would be purified, the carbon dioxide mixed with other chemicals to reproduce oxygen.

The sun would be called upon to supply all power to operate ventilating apparatus and similar auxiliary equipment. The sun's rays would be focussed by reflectors on pipes containing circulating water, the intense heat thus created flashing the water into steam to run small turbo-generators.

It has been suggested that plants be used in the living-quarters to change our expired carbon dioxide into oxygen.

Plants, as we know, keep our terrestrial atmosphere purified by absorbing carbon dioxide from the air and releasing oxygen, the reverse of our body chemistry. By distributing a sufficient number of well-chosen plants throughout the bare and cheerless living quarters of the ship the proponents of this plan believe that not only the air would be kept purified, but also a tie with green earth would be maintained.

1 The necessity for a proper atmospheric pressure within the ship is as important as the presence of air itself. Our bodies at the earth's surface have a pressure upon them continually of nearly fifteen pounds for each square inch of body surface, or a total pressure of some 30,000 pounds. Were this pressure suddenly removed it is probable that our bodies would expand outward like the blowing up of an air-filled balloon This would certainly happen to us if we stepped, without protection, into airless space.

It is to be doubted, except for a journey of several years, whether this plan would work. The weight of plants necessary to supply the requisite oxygen would be greatly in excess of that of any purifying apparatus. But the idea is sentimentally attractive, and offers a field of research in which experimenters may well make interesting contributions to an important problem in this new science. The type of ventilation that is finally adopted will, of course, be that which requires the least total weight of equipment and apparatus.

The maintenance of a comfortable temperature within the ship admits also of practical solution. A highly polished surface on the side of the ship facing the sun would probably reflect away seventy-five percent or more of the heat received. The remainder that is absorbed would be conducted through the metal hull and warm it equally. Since a vacuum can be maintained between the outer and inner hulls, and a thermos-flask effect obtained, as much or as little of the absorbed heat as desired would be allowed to penetrate into the ship. By coating the part of the side of the ship turned away from the sun a dull black a maximum of heat will be absorbed from the reflected sunlight of the earth or moon.

We know that in boiler-rooms of modern power plants a comfortable room temperature is maintained even when, beyond a few inches of steel, there is a roaring furnace. Similarly with a terrific heat on one side and heatless emptiness on the other; our explorers, encased in their thermos-flask, would have little fear of being either roasted or frozen in interplanetary space.

The ultra-violet rays of the sun, so necessary to good health, can be obtained directly. A porthole facing the sun, constructed of quartz (which will admit ultra-violet rays) and of such a thickness that only the desired amount will pass, should provide our explorers with all the advantages of terrestrial sunlight.

Reviewing the barriers to the space flight, we find many difficulties to be overcome, none of which actually prohibit the flight. Minor obstacles in the construction of the ship will be numerous, and years of trial-and-error experimentation will be necessary. All astronauts understand this, and are willing to reckon with it. They see the space flight blocked by barriers that they know will be one day inevitably conquered.

IV

There is still another problem of an interplanetary journey that requires some discussion, and that is the charting of a space-ship to its destination.

Interplanetary navigation or astrogation, as it has been called, is in theory an exact science, yet one requiring the highest order of technical skill. The ship, once it leaves the earth's influence, must follow a course as definite almost as that outlined for a train by its steel lines. And, although slight deviations from this rigidly fixed path in a trip from earth to the moon may not be attended by more than a useless expenditure of fuel; during a journey to Mars or Venus, appreciable changes from the course may lead to disaster, with the plunging of the space-ship into the gigantic furnace of the sun, or into the far reaches of interstellar space.

The planetary bodies maintain their invariable orbits about the sun by a balance between the centrifugal force of their speed through space and the sun's gravitational pull. To change their orbits they must change their speed, so that the balance will not be broken. The speed of a solar satellite through space is thus determined by its distance from the sun.

As we go outward from the sun to Venus, the earth, and Mars, the orbital periods of these bodies — that is, their year — becomes greater and their orbital speed less.[1]

In order for our space-ship to balance in space the onerous pull of the sun it must at all times pursue a definite orbit with relation to the sun; and its orbital speed must at each moment be equal to that which a planetary body would have at that particular distance.

But even on a Lunar journey we must consider this fundamental in all interplanetary navigation. The moon is continually moving in its orbit about the earth, and to hit the moon means the exercise of the same tactics as would be used by a hunter in shooting at a flying bird. The shot must be aimed ahead; the path of the ship through the heavens must be directed to the point where the moon will be at the journey's end.

The only sign-posts for the ship's course are the fixed stars. We know that the moon moves through the heavens against a background of known stellar constellations. If the journey is to take forty-eight hours the place in the heavens to be occupied by the moon at the end of that time is noted, and the ship directed to it.

1 The orbital periods of the three planets mentioned, for example, are respectively 225, 365, and 687 earth days, and their speeds 22, 18, and 15 miles per second. The earth and moon, for our purposes, may be looked upon as a single body, for the moon, revolving about the earth, partakes of its velocity round the sun. A ship travelling from the earth to the moon would therefore hold the earth's velocity about the sun, making the requirements for a Lunar journey far simpler than for one to Mars or Venus.

To aim too far ahead of the moon would mean a premature arrival, requiring a useless expenditure of time and fuel, as well as intricate calculations for swerving in an extended orbit, to meet the tardy satellite. And such an expenditure cannot be permitted if the journey, as it seems likely at present, must be planned with an eye to minimum fuel consumption.

To aim "behind" the moon means arriving at the rendezvous too late, and settling down to a stern chase to catch up with our speeding satellite. Even on a Lunar journey the course of the space-ship is thus rigidly determined.

V

When we plan a journey to Mars or Venus the really precarious situations arise. We have seen that in order to arrive safely we must aim ahead of the present position of the planet of our destination. Furthermore, at the end of the journey we must have a speed, relative to the sun, equal to that of the planet we are trying to reach.

Though we leave the earth with its acquired velocity of eighteen miles per second relative to the sun, when we reach Venus our speed relative to the sun must be twenty-two miles per second, and at the end of a Martian journey it should be only fifteen miles per second.

And to determine whether our course, speed, and direction are correct at any time we must know the exact attraction of the earth, sun, and the planet of our destination exercised on the ship.

The requirements of speed and time of arrival can leave little to chance or experiment. The navigator of the space-ship must know that his craft is capable of the journey proposed; and, further, he must know at each instant during the trip whether it is on its calculated path.

Certainly a new science of navigation must be developed to meet these terribly rigid and novel conditions. Just as two-dimensional navigation on the earth's surface gave way to navigation when men attempted to travel through the air, so in interplanetary travel we must develop an exact science of three-dimensional astrogation through the heavens.[1]

1 This is especially true since the planes in which the planetary orbits lie are not the same. The space-ship must not only travel "forward" and to the "left" or "right" (using these terms in a general descriptive sense), but it must also travel "up" or "down" to get into the new planet's orbit. Although at a single instant the earth may be 92,000,000 and Venus 67,0000,000 miles from the sun, the distance between the two planets will not be their difference or 25,000,000 miles; but, due to the different planes in which they lie, may be 25,300,000 miles. In astrogation this difference of more than a quarter of a million miles is quite vital.

The situation is akin, in greater degree, to the piloting of an aeroplane through impenetrable fog. The pilot must be able to determine not only his latitude and longitude, to know whether he is safely on his course, but also his height above the ground.

The ground in the interplanetary flight is the plane of the orbit of the planet to which he is travelling.

A further condition limits the freedom of a spatial navigator. He can commence his flight only at certain definite times if he expects to win through.

The distances between the planets are, furthermore, constantly changing, being at a minimum when they are on the same side of the sun and in a direct line with it, and greatest when they are in a line with, but on opposite sides of, the sun. Thus the earth-Venus distance may be as little as 26,000,000 and as great as 160,000,000 miles, while our distance from Mars varies from 50,000,000 to 235,000,000 miles.

Naturally, then, the journey should be arranged for a time when the earth and the planet to be visited are closest to each other. These periods occur for the earth and Venus every nineteen and a half months, and for the earth and Mars every twenty two and a half months.[1]

These favorable times must be rigidly observed not only in the planning of the outward journey, but also for the return. Unless the stay on the alien planet is to be for a few weeks or less, the careless explorer will find, if he has been leisurely in making up his mind to return, that his distance from earth has so increased that he is stranded until the next period of minimum distance, possibly a year and a half away. Each day will separate farther the planet from the earth, until at last the sun itself will stand between the explorer and his home.

Let us see how these rules must be worked out in practice.

According to Robert Esnault-Pelterie,[2] a journey to Venus should take some forty-eight days. If the trip is to be made on a course requiring the least distance

1 When Venus and earth are in a line with the sun and on the same side they are said to be in "inferior conjunction." "Superior conjunction" occurs when the sun is between them. With Mars the time of shortest distance is called the "opposition." At "inferior conjunction" with Venus that body is seen by us near the Earth's disk; at our opposition with Mars the Martians will see the earth nearing the sun's face while we still see Mars in a telescope as a sun-illuminated disk.

2 Robert Esnault-Pelterie L'Astronautique.

the space-ship must start about forty-eight days before "inferior conjunction." At that time Venus will be on the left of the sun, as seen from the Northern Hemisphere of the earth, a brilliant star, visible in the west just after the sun has set.

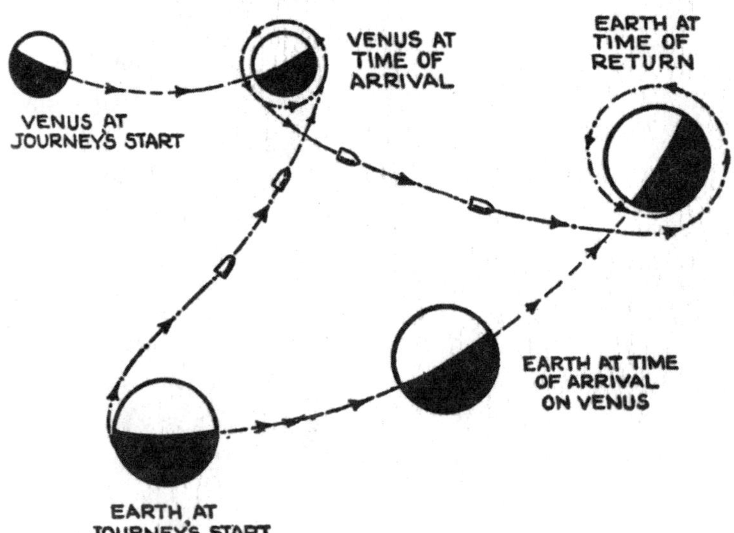

FIG. 6. THE COURSE OF A SHIP ON A FLIGHT
TO VENUS, AND THE RETURN TO EARTH

Our astrogators will employ mathematical astronomers to work out painstakingly a series of charts, plotting a course through the heavens that will bring them to where Venus will be forty-eight days after the journey is begun.[1]

At the journey's start the ship must first rise from the earth and acquire a speed of seven miles per second (relative to the earth) in order to escape from the earth's attraction. Only then can it be set on the predetermined course toward the orbit of Venus. It still has the earth's velocity of eighteen miles per second relative to the sun[2] and this velocity must be increased so that its speed relative

1 This path will be a fixed curve, which, if continued, would take the space-ship around the sun and bring it back to its starting point.

2 The ship will also have a speed of seven miles a second with relation to the earth. If the shot is aimed from the earth in the direction of the earth's orbital movement the ship will be moving through space at eighteen plus seven miles a second. If it is aimed in a direction opposite to the earth's motion it will have a speed through space of eighteen minus seven, or eleven miles a second.

to the sun will be twenty-two miles per second as its path converges with Venus forty-eight days later. Speed, direction, time, and place are all fixed in advance for the explorer by the iron laws of celestial mechanics.

If all these conditions are maintained the ship will approach the vicinity of Venus moving at the same speed as the planet, and in the same direction. The ship can be carefully drawn into the planet's gravitational field, and can circle the world or land upon it.

Surely a complete faith in the laws of mathematics and astronomy must be the first requisite of the interplanetary astrogator. For as his little craft is lost in the immensities of space, and he sees the fiery sun on his left, waiting to snatch him, Venus, lost in the sun's glare, will not even be in sight. He must realize then that he is staking his life and that of his crew on the knowledge that Venus will be at an appointed place to meet him. He must indeed have a sublime faith in the inflexibility of heavenly mechanics and of the power of his craft!

He will find, as day after day passes in the changeless skies, that doubts will arise as to whether Venus will meet him at all. The earth will have shrunk to a ball, then a disk, and finally to only a point of light in the heavens The emptiness will seem infinite. The astrogator will take observation after observation of the stars, check and recheck his speed and course, and wait feverishly for the first glimpse of emerging Venus. Doubt will change to fear, and then to terror, as the weeks bring only increasing loneliness and monotony. A hundred times he will believe himself lost, and, like the first navigator of the Atlantic, he will be besieged by a panic-stricken crew demanding that he turn back before they perish miserably in this infinite space.

He will have to fight these things, as well as a doubt of his sanity as the terrible stillness of space envelops him. But if he remains steadfast he will at last find the justification of his faith in the approach of Venus in the skies. He will see it grow day by day, as their paths gradually converge, from a tiny to a sizeable disk, then into a gradually expanding ball. From an unblinking star it will become a round, fair young world. And as at last he adjusts his course to circle the planet or to land on its surface he will find that he has acquired from this experience the steel and strength of a cosmic faith.

CHAPTER XIV

NEW ENGINES AND OTHER POWER

I

MAN, faced by obstacles in nature, has an engaging way of combining cunning with scientifically trained intellect to attain his ends. History has demonstrated this again and again, in the invention of the aeroplane, wireless, television, and even of such (to us) simple devices as steam-engines and power-driven boats. Seemingly insuperable obstacles are often to inventive minds merely challenges to extraordinary accomplishment. There is reason to believe that the solution of the problems of space flight, now so earnestly desired by persons of scientific training, will be brought about by the application of hitherto untried principles or the practical use of suggestions which to the conservatively inclined may now seem fanciful.

This chapter is an attempt to sketch, as far as science can guide us, the fields of possible power and energy, to survey what forces man can bring under his control for use on a space flight.

Admittedly, among the barriers to the conquest of space that of fuel is the most serious and pressing. If the fuels we now possess and the engines we control are impractical or inadequate we must try to discover more fitting ones and to adapt them to our uses. Certainly in a world in which energy is everywhere freed in such prodigal streams there are riches that man can make his own. In the heart of matter, in the sun, and in the movements of the heavenly bodies there reside possibilities to satisfy the most greedy. The nature of these forces, and what can be done to place them under our control, will form the basis of this discussion.

Thc suggestions that follow may none of them ultimately prove worth consideration. It may well be that they will require more money, energy, and apparatus than their value warrants. But they are offered as hints to fields of research that may yield fruitful results. Only experiment can determine their value.

II

The voyage of Terra was assumed to take place as the moon approached the "half full," or, as the astronomers say, "at quadrature." The course of the space-ship was then neither directly toward nor away from the sun.

The speed attained by the ship removed it from the earth's to the moon's influence, so that it crossed the neutral point of their mutual gravitation (like coasting over the summit of a hill) with velocity to spare. The dividing line was found to exist 220,000 miles from the earth.

It is suggested now that the journey might instead be made at the new moon, in order to use the attraction of the sun to the greatest advantage. The moon is then between the earth and sun (though not necessarily eclipsing it), and the moon and sun can exert their joint attractions to pull the ship from the earth's grip.

General calculations show that then, instead of making it necessary to travel 220,000 miles to become free of the earth, the space-ship, with the sun's assistance, will reach the neutral point at 160,000 miles from the earth. If this is true, then a smaller velocity, with a consequent saving in fuel, should suffice to enable the ship to reach the moon.

The net gain from the manoeuvre would be, of course, not so great as it first appears. Since the ship enters the moon's field much earlier than before, its speed as it falls toward the moon would ultimately attain a greater value, and more power would be needed to check the speed so that a Lunar landing might be made.

There would still be a decided saving, however, for the moon's attraction is only one-eightieth that of the earth's at an equal distance.

For the return to earth the sun's attraction could also be called into play if our explorers could survive on the moon for two weeks. The moon would then be full, as seen from earth, and the sun and earth would jointly oppose the moon's attraction for the flier. The ship could escape, falling back into the earth's influence, with a proportionately smaller consumption of fuel. The sun's gravitational pull is therefore one force that can theoretically be utilized to advantage.

III

Of all the forms of energy thus far considered for use in the space flight electricity has been notably neglected. Yet in electricity we have the most compact and easily applied source of power known. Let us consider how electricity might be employed to lift a space-ship from the earth.

When Terra began her journey she used her own energy to start herself over the inclined rails for a leap into space. There is no valid reason why auxiliary aid cannot be called into play not only to supply this initial impetus, but also to give the ship a generous start upward. For this purpose the building of a magnetic railway has been proposed.

Suppose, as has been suggested, that the rails were inclined up a gentle slope three or four miles in length. Along them would be placid powerful electromagnets operated by an automatic circuit, so that their polarity, or magnetic effect, could be reversed at any moment. Similar magnets on the car which carried the ship would work with those on the rails in attraction or repulsion.

The ship would be pushed along the line by a locomotive engine until a speed of some fifty or sixty miles an hour had been reached. At that speed the engine would be automatically uncoupled and the ship sent flying over the rails into the field of attraction of the first pair of magnets. These would pull it onward and accelerate the speed. As the ship passed them their polarity would be reversed. They would now repel it — pushing it with increasing speed, to the next pair. These would likewise attract and then repel it by turn. Gathering speed at every instant, the ship would sing over the rails, its path inclining upward more and more, leaping at last into space at a speed of hundreds of miles per hour.

Since the impulse given to the ship could be increased by increasing the size of the magnets a tremendous initial velocity might thus be provided without the use of a pound of the spacecraft's precious fuel.

The value of this suggestion,[1] however, is not to be pressed too strongly. The cost of the railway with its expensive magnets would undoubtedly be great, and the total momentum imparted to the ship would necessarily be only a small fraction of the total necessary for the flight.

But the magnetic railway does indicate how an auxiliary aid might be used to ease the burden of the ship's fuel requirements. Other suggestions, ranging from the use of a gigantic wheel to fling the ship into space by centrifugal force at a terrible speed, to the utilization of gravity itself, by the dropping of a counterweight to pull the ship up a mountain-side, all offer fanciful approaches to a solution of the starting problem. But despite the ponderousness and awkwardness of the methods suggested they have the virtue of showing that we have not yet begun to command the forces that are everywhere about us. It is probable that the number of auxiliary engines that could be devised are endless. In one of them may be found the germ of a truly startling discovery.

1 Described quite vividly in a story, "The Moon Conquerors" by R.H. Romans, published in the Wonder Stories Quarterly Magazine, Autumn 1929.

IV

The question of energies and fuels, as contrasted with engines, that might be used for the space flight opens up one of the most fascinating fields in all science. Modern physicists believe that if we could plumb the heart of matter and seize upon its essential physical nature we might be able to turn tremendous stores of energy to our own use.

Perhaps the closest approach to this El dorado has been provided by that modern wizard of chemistry Dr. Irving Langmuir, in his discovery and isolation of "atomic hydrogen."

We know that hydrogen as found in nature is in the molecular form: two atoms, like twin cherries, forming a single unit.

Langmuir was able to break down the two-atom hydrogen molecule — separate the twins, as it were and keep the element in its highly unstable monatomic form for a time.

The calculations of Robert Esnault-Pelterie on Langmuir's discovery led to the conclusion that liquefied, mon-atomic hydrogen might yield on combustion ten times as much energy as the best smokeless powder, and over three times as much as the powerful hydrogen and oxygen mixture!

Here certainly lies a possible solution to the fuel problem. Using such a fuel in the step rocket would permit a considerably greater pay load for an interplanetary ship. It would even make a landing on the moon feasible, by bringing the fuel necessary for an interplanetary expedition down to more practicable levels.

But two factors bar the realization of the easy solution offered by atomic hydrogen. In the first place it cannot as yet be liquefied, nor can it be kept for long in its unstable state. Inevitably it changes to the more stable — and less powerful — molecular diatomic form.

Secondly, the utilization of the fuel is accompanied by the liberation of so much heat that the resulting gases might well reach a temperature of 18,000° Centigrade. No metal, or any substance possible for use in a rocket, could stand this heat.

Pelterie himself feels that, however desirable it may be, we must regretfully turn away from atomic hydrogen as a fuel — at least, until its nature is better known and its action better controlled. But the terrific energy it contains causes the eye of rocket-builders to rest upon it with continued fascination. It offers a goal worthy of the efforts of other Langmuirs. Perhaps one of them will find a key to unlock the door to its use. If not that, then other compounds — synthetic if necessary will undoubtedly come from the retorts of the chemists to give us in full measure the power we require.

V

We now leave the immediate and presently practicable for the realms of the speculative.

In the field of alluring possibilities resides the hope of extracting the giant stores of energy contained within matter itself. We know that in far-off nebula; matter is constantly being created from energy and destroyed again to become primal energy.

Modern physicists tell us that matter and energy are basically the same. Scratch an atom, and we find locked energy.

The atom is thought to be composed of a nucleus of matter (of infinitesimal size) possessing an electric charge. About it revolves one or more electrons, which are also infinitesimal particles, whose charges are opposite to those of the nucleus. The electrons have tremendous speeds — several thousand miles a second — in their orbits. By these speeds alone (as in the case of the planets in their orbits round the sun) they resist the attraction of the nucleus.

Energy is needed to create these systems and to give the electrons their great speeds. Energy will similarly be released if the atom is destroyed. "This energy in matter," says Sir James Jeans,

> is of an entirely different order from that made available by any other treatment. The combustion of a pound of the best coal in pure oxygen liberates about 1,600,000 foot-pounds of energy;[1] the [atomic] annihilation of a pound of coal liberates 31,300,000,000,000,000 foot-pounds or about 18,000 million times as much. In the ordinary combustion of coal we are merely skimming the topmost cream of the energy contained in coal, with the consequence that 99.999999994 percent of the total weight remains behind in the form of cinders, smoke, and ash.

1 Sir James Jeans, The Universe Around Us (Cambridge University Press). I have here taken the liberty to reduce Sir James Jeans' original figures to foot-pounds of energy per pound of coal. They were originally in ergs of energy per ton of coal.

Annihilation leaves nothing behind; it is a combustion so complete that neither smoke, cinders, nor ash is left. If we on earth burned our coal so completely as this a single pound would keep the whole British nation going for a fortnight — domestic fires, factories, trains, ships, and all; a piece of coal smaller than a pea would take the Mauretania across the Atlantic and back.

This is the pronouncement of one of Britain's most eminent men of science.

His figures mean, in effect, that the utilization of the complete atomic energy of a pound of matter would be sufficient to transport a 500,000-ton spaceship from the earth to the moon and return!

These staggering figures may serve, perhaps, to make the hope of the utilization of atomic energy seem entirely fanciful. They imply riches too great; they escape our sense of reality. Not only would the release of this energy solve the interplanetary question, but it would also solve the whole problem of power for our economic and social needs, and from that standpoint provide a veritable Utopia.

Yet there is no theoretical reason why this miracle might not come about. The problem has already attracted scientists of the first rank. Among them is Dr. Arthur H. Compton, Nobel prize winner in physics and a professor of the University of Chicago. Dr. Compton, after preliminary studies, has become sufficiently impressed with the possibility of obtaining atomic energy to turn his full energies and his extraordinary gifts to the attempt to discover a method of releasing it.

Were this energy available there would be no thought of the dangers of acceleration, of weightlessness, of maneuvers for landing, of rigid routes or the rationing of food. Ruling the empire of matter, we need take along only a few "atomic bricks," which, by the use of ingenious apparatus, would cause giants of power to spring to our bidding.

Two problems are fundamental in this golden dream of power : How is the energy to be released? and how is it to be used?

To the first question we can turn directly to nature, where, in the radioactive minerals, such as radium and uranium, the process is going on continually. Without aid of any kind, defying, in fact, all attempts to accelerate or to retard their action, these minerals are decomposing, changing ultimately into atoms of lead, and shooting off particles of energy in the form of swiftly moving electrons. And just as gas particles of our ordinary fuels propel a rocket by their ejection, so these alpha particles of radium may some day be ejected from the rocket tubes of an interplanetary ship.

Thus in radium itself we have a source of atomic energy, and no energy is needed to start the decomposition. We learn that in the disintegration of ten pounds of radium, if that much were available, there would be enough energy released to raise a ton and a half of space-ship beyond the limit of the earth's influence.

A difficulty in the utilization of radium, apart from the scarcity of it, lies in the extreme slowness of the process of disintegration. After decomposing for five million years radium will be only partially consumed. We can hardly wait that while! Nevertheless, the action of radium shows what a great store of power lies within the reach of scientists if they can only speed up what is now going on as a fixed natural process.[1]

Radium, however, is not the only substance from which, theoretically, we could obtain the release of atomic energy. It is merely an illustration of atomic disintegration in a simple and universal form.

If we knew the secret, perhaps any element could be decomposed to release its vast store of power for our use. But failing in our search for a method of decomposition, still another way to obtain enormous atomic power has been suggested. It is known as the method of "building up" elements, or the method of taking advantage of "mass defect."

It has been claimed, for example, that the source of cosmic rays is the distant nebulae, or the hot interior of the stars, where these rays are thought to originate during the process of building up atoms of helium, or perhaps nitrogen, from lighter atoms, such as hydrogen.

By celestial book-keeping we learn that fourteen units of hydrogen might form one unit of nitrogen. But of the original weight of hydrogen — say 14.112 pounds — only 14.008 pounds of nitrogen would result.[2] Since neither energy nor matter can be created or destroyed, it is concluded that the .104 pounds that are apparently lost are shot off in the form of radiant energy.

Applying Sir James Jeans' figures for the atomic energy possible of release in such a case, we find that one pound of hydrogen would yield .992 pounds of

1 The announcement in June 1931 of the construction of a 16,000,000 volt Coolidge tube by two German scientists, Dr. F. Lange and Dr. A.Brasch, shows that a solution may not be far away. With their tube the scientists expect to duplicate artificially the action of X-rays, radium emanations, and cosmic rays.

2 The atomic weight of hydrogen is 1.008. Fourteen hydrogen atoms would weigh 14.112 units.

NIGHT VIEW OF A SUCCESSFUL ROCKET SHOT

nitrogen and 246,300,000,000,000 foot-pounds of energy, enough to raise 4000 tons beyond the limit of the earth's influence.

The disintegration of matter or its complementary process, and the consequent release of energy, is thought to be going on continually in the universe, not only in the nebulas, but also in our own sun. Only by the release of such tremendous stores of atomic energy, it is argued, has the sun continued to radiate at its prodigal rate for many millions of years without suffering a great decrease in size. It has often been pointed out that if such disintegration is a natural process there is no reason why man cannot duplicate it in his laboratories.

It is quite possible, of course, that the energy required to tear atoms apart and to rebuild them would not compensate for what we might get. Atoms are probably broken down in great suns only by terrific heat, several millions of degrees and upward, and by the tremendous pressures within their interiors. Such conditions we could hardly duplicate on earth.

Our only success in this field to date has been in shooting electrons produced by high potential X-ray tubes against gases, and knocking occasional electrons from their atoms. Rutherford, in his studies, shot alpha particles of radium through nitrogen gas and found that the radium particles, travelling at several thousand miles a second, were able to disintegrate pieces of matter to form atoms of hydrogen. But such atomic transformations, suggestive as they may be, do not produce energy; on the contrary, they require the expenditure of great quantities of it.

It has been noted with some justice that the idea of atomic energy carries its own contradiction. It is all very well, say the skeptics, to speak of the disintegration of material in great solar furnaces, for no container is needed to hold the material or the released energy; the whole sun is blazing away at a staggering temperature.

But the release of atomic energy on earth, they state, would be impossible, for there is no material that could withstand the resulting heat and pressure. It would be equivalent to the old story of discovering a universal solvent — if its power were as specified there would be no container in which it could be used or kept.

Much can be said on both sides of this argument, but it would all be theoretical and hypothetical. The dangers inherent in the release of such tremendous blocks of energy, however, bring up the important question of how it could possibly be used. Applied to the rocket, particles of swiftly moving matter would have to be shot from the exhaust tubes, the reaction sending the rocket on its way.

In the disintegration of radium this process is produced naturally. Pieces of radium could be placed in a special disintegration chamber and the alpha particles, shooting out of an exhaust at a speed of several thousand miles per second, might perhaps produce an enormous recoil. The use of such energy, however, would be very wasteful. An expulsion speed of 2000 miles a second, to be used on a seven miles-a-second ship, would be productive of very low efficiency.

In the complete disintegration of matter the problem is even more indefinite. We are not at all certain of the form the released energy would take. It is quite possible that it might be a variety of cosmic ray, in which case its action, perhaps, could not be controlled. The rays might shoot out in every direction, through the hull of the ship, into the passengers' quarters, through the control chambers, and might prove to be like the genie that rose from the bottle suddenly to overwhelm his discoverer.

The problem, however, is a fascinating one, despite its present impractical aspects. It is foolish to say with certainty whether or not such control over matter can be attained. On the other hand, until the time comes when its use is proved definitely impossible the subject will attract a growing number of keen minds. The ultimate answer resides in the future.

VII

The astronaut seeking for the release of the spaceship from the earth's pull must think occasionally:

"If only the gravitational attraction of the planets were to be nullified or banished for a few minutes, what a vision of interplanetary conquest would open!"

The road to the planets would then be a light, joyous climb. Weight having ceased for us at the earth's surface, the centrifugal force would send the ship soaring into space. We could travel from planet to planet with careless abandon. Rockets using only a few pounds of fuel to accelerate the speed, for steering, and to check the speed occasionally, would be sufficient for a trip to Mars or Venus. Rigidly planned routes would not need to be thought out for this delightful journey — it would be a carefree jaunt for crew and navigator alike!

This thought is by no means new. Astor and Wells used it to good effect in their stories of interplanetary voyages.

Both "discovered" substances used to cover their ships, to shield them from all gravitational influence as we use shields against heat and light. Their ships were freed of the attractions of any planetary bodies, and rose lightly from the earth to move in leisurely courses to their heavenly destinations.

The people of Astor's and Wells' day, however, treated these stories simply as fantasies, for it was generally accepted that what Newton had postulated was true: that wherever matter existed there must be gravitation, and that gravitation was universal in its range.

The recent pronouncements of Einstein, however, lend a more serious note to the general idea of gravity nullification. Gravitation and electromagnetism are sister and brother, said Einstein, therefore they should have a common parentage, and that parentage he is attempting to discover. From the laws of the one might be found the key to the nature of the other. Matter which causes gravitation is only electromagnetic energy at bottom; thus implying that gravitation is only a more elusive form of electromagnetic attraction.

From these hypotheses fruitful speculations arise. Can gravitation be overcome by electromagnetic means, and a shield really devised for it, just as we now have shields for the other electromagnetic forms of energy — heat, light, wireless, ultra-violet rays, etc.? In other words, are gravitational phenomena only other sections of the electromagnetic spectrum?

Those who scorn the idea of gravity nullification state that it is contrary to all the laws of nature. If, they say with justice, we could devise a shield against gravitation, and therefore raise a weight above the earth with little expenditure of energy, we could remove the shield and allow the weight to drop back to earth to do considerable work. We thus cheat nature's "conservation of energy" by obtaining work for nothing — almost as good, in fact, as perpetual motion.

If mere shields of some "unknown" substance, such as Wells and Astor used, were the whole story of gravity nullification, critics would be justified in calling it a fool's dream. But apart from the realms of fiction it is hard to conceive of serious men of science believing in such easy trickery over nature. No mere shield or blanket alone can overcome universal gravitation.

Those who believe in it at all propose that, since gravitation is a form of electromagnetic energy, energy can be used to overcome it. A ship would be "charged" with sufficient electromagnetic energy to give it repulsion instead of attraction for the earth. The work done in lifting the ship would be all "stored" as repulsion energy in the material of the ship. The ship would, in a sense, be magnetized with a polarity opposite to the gravitational magnetism of the earth. The total amount of energy required for this magnetization would, without doubt, be equal to that necessary to lift the ship from the earth without such a device. Thus Nature's laws would always be satisfied.

The advantage of such a proposal, visionary as it is, lies in the fact that all such energy would require little weight. When the ship was ready to begin its journey it would have within itself the power to lift it away from the earth, and perhaps away from the other planets. The journey would be safe, and comfortable, and one could depend upon the ship at all times.

Although no practical demonstration of these glowing fantasies has been as yet obtained, and no actual programme for their use outlined, a few rays of hope do exist. Engineers of the Bell Telephone Laboratories devised a metal bar composed of an alloy of nickel and cobalt which, when magnetized, resisted the attraction of another bar magnet, and remained suspended above it, in partial defiance of gravitation. The engineers, however, have not been energetic in pursuing their success, and it is probable that no promise of any practical achievement was offered to them.

But advocates of the gravity nullification method of interplanetary flight believe that what can be done in a small and unimportant way can also, with further research, be accomplished upon a larger and more pretentious scale.

The search of Einstein for equations that will at once explain all forms of energy and their interrelation opens vistas unlimited in extent. Even should Einstein fail at this gigantic task there are dozens of other scientific men ready to carry on his work. The search for the reality of force, energy, and matter has never been so keen as to-day, and its promise for space flight never so bright.

In the sun, in radium, in all elementary matter, in electricity, are forces ready to do man's bidding if he is wise enough to learn the secret of their control. None of them may be available before the first space flight is made, for the true interplanetary enthusiasts do not wait on golden dreams of discovery. They work with the tools they have-using fuels over which control is possible, extracting from them the last foot-pound of available power.

Leaving this question, therefore, we must realize that, although the future is rich in promise, we must for the time being turn resolutely away to face present realities.

CHAPTER XV

NEW WORLDS

I

THE question of overcoming the mechanical barriers to the space flight leads naturally to a discussion of what promises and rewards astronauts can offer to a practical world.

The meteorological rocket has been designed only to penetrate our atmosphere and to return to earth with its automatic record of the phenomena of space. The interplanetary flight, with its immeasurably wider implications, presupposes a landing upon other worlds, where daring astronauts may not only explore and have their fill of adventure, but also observe and record information of extreme value to science.

The realization of these rewards therefore must be based on our ability to make a peaceful conquest of our sister worlds. Should they prove so inhospitable, for instance, that we could not remain for any reasonable time, or only with extreme difficulty, exploration would be impossible and the greater rewards to the explorers for the hardships of their journey would be denied them.

Without a preliminary survey of conditions on planets to be visited the expedition, having safely escaped the dangers of space, might still come to grief upon the barren or frozen surface of an alien world, or be engulfed in poisonous gases, or destroyed by hostile life forms.

Other questions also present themselves, particularly of the actual presence of any life forms on other planets. This question, which has been hotly debated by astronomers, philosophers, biologists, and even theologians for many years is of the greatest interest to our explorers. A planet which perhaps supports a cycle of life forms offers a rich field of discovery for pioneering scientists. Such a world might also be fraught with dangers even worse than those imposed by the harshness of their physical natures. It would be foolish for astronauts to set out on a journey to explore any of the planets without carefully surveying these possibilities and preparing for them.

The actual determination of planetary conditions, and whether or not there are living creatures upon our sister worlds, must of course await the space flight itself. But fortunately we can, by the use of modern astronomical instruments, attain a rough approximation of the physical conditions of the planets, to determine whether in those circumstances life might exist.

What, in the first instance, are the conditions necessary on other worlds for the support of animal and vegetable life as we know it?

Dr. Harlow Shapley of the Harvard Observatory has outlined seven such conditions. To him they are the circumstances necessary for the initial production of life on a formerly barren world. But they may be used by us as indices of the continuance of such life to the present, and serve also as an index of the possibility of man's existence on them.

Dr. Shapley outlines these conditions

1 . The sun's radiation must be constant in quality and quantity over a considerable period of time.

2. The distance of the planet from the sun must be such as to permit the existence of water in a liquid form.

3. The orbit should be approximately circular, so that its solar distance shall not vary greatly.

4. It must have a satisfactory rotational period, so as to obtain endurable day and night.

5. The axis of rotation must be suitably inclined.

6. The mass of the planet must not be too small or too large.

7. The chemical constitution of the planet's covering of land and water must conform to definite prescriptions for life of terrestrial kind.

Most of these requirements will, on reflection, appear self-evident. It must be realized that terrestrial animal and vegetable life consists of delicate, unstable compounds of organic molecules. They require definite physical conditions for the maintenance of their structure against breakdown and decay. The conditions must not only be right, but they must not vary too widely — that is, their variation must not be too great from our point of view. Animal and vegetable life such as we know can exist only within a narrow band of the possible conditions in the cosmos, and would require only a minor change to wipe it out entirely.

II

The first requirement, therefore, is that variation in the heat, light, and other life-giving rays of the sun be within very narrow limits. An organism might have come into being and have existed at the frightfully high (to us) temperature of 55° Centigrade (131° Fahrenheit). But should the prevailing temperature drop to 30° Centigrade (a change that is infinitesimal in the cosmic scale) it is quite possible that the organism would soon die. This holds true similarly for any changes in the other solar rays necessary for the maintenance of life.

We know further, from our terrestrial experience, that the existence of life forms, reaching up to man himself, demands a definite as well as a stable range of temperatures. Such limits can be placed roughly between 125° and -60° Fahrenheit. Above this limit — for human beings at least — blood would probably coagulate. Below it we should undoubtedly freeze. The necessities for animal and plant life which might be assumed to exist on other worlds are unknown to us, but assuming them to be upon the same general scale, we may see how conditions on the solar planets compare with our own.

The latest and most complete answer to the question of planetary temperatures has been supplied by the studies of Dr. Edison Pettit and Dr. Seth B. Nicholson of the Mount Wilson Observatory. Using a thermocouple (a delicate instrument for recording the effect of heat) on the principal focus of the great hundred-inch telescope at Mount Wilson Observatory, they recorded the heat reflected to the earth from the moon and the other planets at various times of the year. For this work they had the best conditions and the most sensitive heat-recording instruments known.

The moon, they found, had temperatures ranging from 243° Fahrenheit below zero on the dark side to 216° Fahrenheit above zero on the sunny side. Under such temperatures, entirely unknown on the earth, no animal or vegetable life that we know of could exist. On the basis of temperature alone we must conclude that the moon is lifeless, and that man could exist on its surface only with extreme difficulty, excepting in the narrow band of its surface that is between day and night, and where a comfortable summer heat exists.

Mercury, the planet closest to the sun, has temperatures reaching 720°,making it a sun-parched, barren world. The recorded night temperature of Venus, according to Pettit and Nicholson, is 23° below zero.[1] This, it must be conceded, is not the

1 When conditions are favorable for observing Venus we can see only its darkened side. These temperatures are all on the Fahrenheit scale.

surface temperature of the planet, for its thick cloud layers prevent the observation of surface heat radiations. It is possibly true that the surface night temperature on this basis may well be ten or twenty or even fifty degrees above zero, comparable to conditions in our temperate zones.

The temperature of Mars varies greatly not only each day, but by seasons. When the planet is closest to the sun its noon temperature is 72° above zero — a balmy, healthful level. When it is farthest from the sun its noon temperature descends as low as 40° below zero, a chill equal to that of the sharpest mid-winter days in Central Canada. Our investigators believe also that the night temperatures of Mars the year round must be below zero; at sunrise and at sunset just above zero.

The climatic conditions on Mars, one and a half times as far away from the sun as the earth, are undoubtedly rigorous, and hardy animals and plants would be necessary to endure them. But these temperatures do not lead us to believe that life could not or does not exist there.

The planets outward from the sun, beyond Mars, are believed to be extremely cold. The temperature of the outer layers of the giant Jupiter is placed at 216° below zero, while the temperatures of the still more remote planets, Saturn, Neptune, and Uranus, are believed to be as low or even lower. These figures, however, are not definitely established. It is possible that we have recorded only the temperatures of the extreme outer layers of their massive clouds.

If a Jovian were to make temperature measurements of the earth he might not know that he was actually measuring the temperature of the upper limits of our atmosphere, and he might conclude that the earth was too cold to support life of any kind. He might conclude also, as we shall see, that the earth's upper atmosphere, being composed mainly of hydrogen and helium, could support no life, and was therefore desolate.

We must admit that our studies of the vapor-masked major planets may be likewise erroneous. Since these worlds are so massive they may have retained a great deal of inner heat, which may not be evident to our instruments. All we can bring in support of the evidence of their extreme cold is their location — so far from the sun that they receive only a fraction of the heat received by the earth.

Mars and Venus and a narrow band on the moon alone indicate that they might be homes for life forms, and places where man might exist in reasonable comfort.

III

The second of Dr. Shapley's conditions, the distance of the planet from the sun for the maintenance of water in liquid form, is also a question of temperature. Liquid water is necessary to all plant and animal life that we know, and water can exist as a liquid only between 32° and 212° Fahrenheit. On Mercury, therefore, whatever water might have existed would have gone up in steam. On Jupiter and the other presumably cold planets all water would have been frozen, while on the moon it would alternate between ice and steam as each side is exposed to and removed from the sun's rays.

On Mars and Venus, however, liquid water could be maintained during a reasonable part of the year under natural conditions.

The third condition also is a temperature problem that of variations in planetary orbit. Part of the rigors of life on Mars would spring from the fact that the orbit is decidedly elliptical, its distance from the sun varying between 128,000,000 and 154,000,000 miles. Naturally, at the farther distance it receives 20 percent less heat than at its closer approach, causing by this factor alone great variations in temperature.

The orbits of Venus and the major planets are practically circular, but that of Mercury has the greatest eccentricity, for the planet may be either 28,000,000 or 43,000,000 miles from the sun, causing a range of temperature from 300° to 700°. This variation serves but to emphasize the utter unfitness of Mercury as a place for any life such as we have experience with.

The necessity of the fourth condition — an alternation of night and day within reasonable periods involves the maintenance of a more or less equable temperature range. The rotation of a planet so that each face is turned toward and away from the sun will serve alternately to warm and to cool it, providing a check against the building up of extremes of heat or cold.

No definite rule can be advanced to indicate what the rotational period of a planet should be. It will depend upon the local conditions. We know that the moon's extremes of temperatures are reached within a few minutes of the coming of the two-week day or night, due, as we shall see, to the absence of an atmosphere. It is probable, therefore, that the lengthening or shortening of the Lunar rotational period would have little effect upon its temperature range. That is also true of Mercury, whose period of rotation is eighty-eight days.

The length of the Venusian day has never been determined, its cloud layers preventing the necessary observations of its surface markings. Estimates of its periods range from one day, like the earth's, to a full Venusian year of 225 days. It is probable that its period lies somewhere in between, the circumstantial evidence being that it is somewhat longer than the earth day. It is hardly to be believed that it is long enough for the temperatures to mount up so excessively as to bar the presence of organic life.

The period of Mars, equal within minutes to that of the earth, is perhaps too long. If its nights were shorter the cold side would have less time for its heat to be radiated away.

The rotational periods of the major planets are excessively short, being between nine and eleven hours. This fact, together with their great bulk, makes the speed of rotation on their surfaces terrific. The sun and stars would appear to dash across the sky and disappear in five or six hours. Day and night would follow each other with dizzying succession. The cloud layers would be heated and cooled swiftly, and the surface would be swept by gigantic storms in which frail life forms could not exist.

In conclusion, although the period of Venus is unknown and that of Mars is a trifle short, these planets still remain as possible candidates for the abode of life.

IV

The desirability of an inclination of the axis of the planet aims at an orderly succession of seasons. It is because the Poles of the earth are pointed alternately toward and away from the sun, due to the inclination of the earth's axis, that we have our cycle of the seasons.

It is not to be doubted that life flourishes more vigorously under changing climatic conditions, provided they are not too extreme. But it is hard to see any absolute necessity for seasons. Although the temperate zone of the earth has produced more hardy plants than the tropics, life in the tropics is certainly more abundant and diversified, and the indications point to a tropical cradle for all life.

On Mars the changes of the seasons should be of aid to life. The melting of the supposed polar ice-caps in spring and summer would cause floods of water over a part of the Martian surface to give abundance to the rainless Martian deserts. The writer, however, can see no general necessity for changes of seasons, and especially none to compare in importance with the sixth condition, dealing with the mass of a planet.

The mass, and therefore the gravitational force, that a planet exerts upon bodies on its surface will determine, in the first instance, the presence of an atmosphere of gases necessary to life. An atmosphere not only gives air for plants and animals to breathe, but also acts as a modifier of extremes of heat and cold, and shields the world from dangerous bombardments of meteors and cosmic rays.

The moon, state modern astronomers, can hold no atmosphere, for its gravitational pull is insufficient to oppose the velocity of gas molecules away from its surface. Thus the moon, because of its small bulk, is airless and, considering again terrestrial forms, lifeless. And its extremes of temperature may be accounted for by the lack of an atmospheric shield to diffuse the sun's rays and to hold its heat in reserve for the sunlessness of the two-week night.

Because of its smaller bulk, and the fact that its surface gravitation is only about one-third that of the earth, the experimental evidence that the Martian atmosphere is less than 20 per cent as dense as the earth's may be theoretically verified. Observations on Mars indicate that its atmosphere is only one sixth as dense as that at the top of Mount Wilson, and one-half as dense as that at the top of Mount Everest, where oxygen-breathing life cannot live. The water vapor in the Martian atmosphere, necessary to replace the natural evaporation of the body, is but 6 percent of that found on Mount Wilson.

The continuance of life on Mars, from the standpoint of atmosphere, must therefore be very precarious. And it is because of the tenuity of this atmosphere that extremely low temperatures are reached in the Martian night. Existence on Mars would be almost impossible to man unless he were protected by a special space suit covering his entire body, and supplying him with a steady flow of oxygen equal to that afforded on earth.

On Venus we find a world endowed with a bountiful covering of air, which could not only hold all the necessary gases for breathing, but also act as a blanket from the increased heat of the sun caused by its proximity to the luminary.

The cloud layers of Venus have never permitted studies of the nature of the gases at its surface. The upper layers are definitely devoid of oxygen (but so are the earth's upper layers). The gravitation of the planet's surface, only slightly less than that of the earth, is sufficient to hold an atmosphere of the type to support animal and vegetable life.

Of the nature of the cloud layers of the major planets we know little. Atmospheres these massive worlds do have; in fact, it is believed by some astronomers that they are composed principally of gas — having only a small solid core.

The second condition governing the mass of a planet is that it should not cause too great a gravitational force at the surface. Not only would terrestrial life be crushed by the tremendous force of gravity on large worlds, but the atmospheric disturbances would also be on so great a scale that they would be unendurable for man, beast, or plant. The surface gravitation of the minor planets is in every case smaller than that of the earth, with the exception of Venus, which is practically the same. That of the major planets, except Jupiter, is approximately equal to or slightly less than the earth's. But on Jupiter the calculated surface gravitation is twice that of the earth, indicating that a 150-pound man would there feel that he was carrying an extra load of 150 pounds.[1] As the planet most favorable for exploration, apart from Mars and Venus, Jupiter at present must be ruled out through the difficulty that would be encountered by human beings in remaining erect on its surface.

The presence of an atmosphere cannot naturally presuppose a planet's ability to support life. The atmosphere may well contain noxious gases, or poisonous exudations. It may be devoid of oxygen. The nature of the atmosphere must be determined by spectroscopic studies from the earth supplemented by observations made actually there by explorers fully encased in protective suits. The question of planetary atmospheres is covered in part by the seventh of Dr. Shapley's conditions. So far as our knowledge goes only that of Mars definitely contains oxygen and water vapor while the presence of these gases on Venus may be presumed as hopeful enough to warrant an expedition.

The chemical constitution of the land and water is also important. All animal and vegetable life springing from the soil must draw from it the necessary ingredients to feed and to cause growth. A soil with poisonous or destructive elements, such as radioactive compounds or volcanic ash, makes it quite probable that no living thing grows there. The continued existence of all life means a delicate cycle of interchange between the soil and the life that it supports, and a favorable soil constitution is necessary.

No definite pronouncement on the soil of most of the planets can be made. The recent studies of our sister worlds by Dr. B. Lyot at the Paris Observatory indicate that the moon is covered by a thin layer of volcanic ash, the remnant of former volcanic activity that may have caused its craters. Mercury and Mars "seem to have a similar surface." Dr. Lyot, however, is not at all sure of his conclusions regarding the Martian surface; and the surface of Venus, of course, is hidden from view. From this consideration Mars still remains a prospective

1 The planet is assumed to be fairly solid, giving a surface gravitation of about 2.5 times that on the earth's surface. The centrifugal force at Jupiter's surface due to his dizzying speed of rotation naturally reduces the gravitational pull considerably.

candidate for human exploration, but from Venus no answer has yet been obtained.

The nature of the soil and other local conditions on Venus cannot reasonably be discovered without an expedition. If the other conditions on the planet are favorable, these might well be hazarded.

There are other factors more subtle and less capable of observation from afar that may definitely make a planet impossible of exploration, even though it harbors life of its own. The presence of strange or harmful forms of germ life, against which we humans have developed no natural immunity, may strike down our explorers with sudden and virulent diseases. The plant life may be poisonous to the touch, or may exude harmful gases. The animal life may be so large and ferocious, or so small and numerous, that there is no protection against them, and existence on the planet would be filled with unknown terrors and dangers at every moment.

Sifting the results of this brief analysis of our sister planets, we find remaining in the list of the possible worlds for exploration, and the possible existence of natural life forms, only Venus and Mars.[1] Not only is the moon conclusively devoid of life of terrestrial nature, but the existence of our explorers there would be extremely difficult, if not impossible, except near the terminator.[2]

Only by the use of suits, which would encase men as in a thermos-flask, preserving their bodily temperature and supplying them with oxygen, water, and a normal pressure on their bodies, could an exploration be made on the moon.

Professor Willem J. Luyten of Harvard Observatory points out that the moon "has no atmosphere, no sound waves can be propagated, and it is plunged in eternal silence. . . . We see the moon through the eyes of science: dead, dry, and desolate, unapproachable, forbidding, silent and savage, but still beautiful."[3] Yet because the moon is our nearest neighbor it will without doubt be the terminus of the first space journey. Because we are not absolutely convinced that our existence there is impossible we will use all the resources of science to try to make it at least a base or jumping-off point for journeys to other worlds. Only

1 The moons of the major planets have not been considered. for we know absolutely nothing about their surfaces. This is true also of the asteroids and of the planet Pluto.

2 The dividing line between night and day. The terminator moves over its surface with the motion of the moon some 160 Miles a day at the latitude of 45° since lunar gravitation is only one-sixth that of the earth's a fast-moving man might almost keep at a comfortable temperature.

3 Willem J. Luyten, The Pageant of the Stars,

demonstration by sad experience that we have no means to cope with its extremes of temperature will deter men from attempting to explore our satellite.

Venus and Mars are definitely in the category of worlds worthy of our attention. The evidence of general planetary evolution pictures Venus as a fair young world, in the same stage of its evolution as the earth was six to ten million years ago. Her hot dense atmosphere is suggestive of the steam-laden surface of the earth before condensation had taken place to form our great seas and oceans. Her plant and animal life, should such exist, might conceivably be on the general order of, though hardly similar to, that on earth during the Palaeozoic era. Almost unbelievable forms of life may abound there, and if intelligent life exists it will be a life that knows nothing of the stars or planetary systems — for the heavens must be invisible from the surface. Venusians would have continuously above them only the impenetrable clouds.

When we turn to Mars, however, the conditions are almost reversed. Professor Luyten, for instance, asserts that, reviewing the different prerequisites a planet must fulfill in order to allow life comparable to life on earth, we find that the conditions are just met on Mars, but no more than that. If therefore life does exist on Mars it must be under severe conditions that can hardly be compared to the comfortable ones of our own planet.

These words sum up the experimental evidence from our position 40,000,000 miles distant. Life, we see, is possible there, and if we are to believe that Mars is an older world than the earth, and that the thinning atmosphere and the consequent loss of its surface water was progressive and slow enough, the life forms that survived may have evolved to meet the growing rigor of existence. What we find there may be just as well adapted to its natural conditions as our terrestrial plant and animal life are adapted to the earth.

The question of the actual existence or meaning of the Martian canals is a somewhat academic subject.

The violent divergence of opinion by authorities precludes any intelligent opinion by laymen. The maze of canals seen by Schiaparelli and viewed again by Lowell for over seventeen years are actually denied existence by some modern astronomers, who say that they vanish entirely when viewed in a large telescope.

Those who admit their existence are divided into camps affirming or denying that they may be sources of irrigation for the dry Martian surface, caused by the melting of the polar snows.

The matter could be discussed, in the manner of the Greek sophists, "unto dawn," and the answer would still lie across the heavens. We know that Mars can support life, and that with proper equipment at their disposal it could support men. And the discovery of what it does at present support will be forthcoming only when the rocket ship at our disposal in theory is built in practice.

We have seen then that three worlds could be visited by our space expedition, with the inference that two — Mars and Venus — may support life of their own of the order of terrestrial life. One more word, however, is necessary to this subject before leaving it, a word of humility and caution.

We have assumed in our analysis that we are searching for life that is of the order of terrestrial forms and subject to the same laws. We have been speaking in brief of life formed principally of complex molecules of carbon, hydrogen, oxygen, and nitrogen — known as protoplasm. We must recognize, however, that the formation of this life may, in fact, be of exclusively terrestrial origin. Although on earth, because of its peculiar properties, carbon forms the basis of these molecules, and all life is limited by the conditions under which they can exist and evolve, we have no warrant for limiting life on other worlds to a similar construction. As Camille Flammarion suggests, we should not be like the fish who assumes that no living thing can exist outside of water.

The possibility of chemical compounds that show the ear-marks of life — growing, reproducing, feeding, and betraying irritability — are varied beyond our imagining. There is a possibility that sulphur and silicon and other elements can well serve for the bases of families of life forms. When, therefore, we reject the impossibility of life on Mercury, the moon, and the major planets we must make the reservation that life obeying other laws and flourishing under conditions hostile to terrestrial life may exist.

The discovery of such non-protoplasmic life would be, in fact, among the great triumphs of the space flight, and would extend vastly the range, scope, and depth of our knowledge of all life.

CHAPTER XVI

PROMISES AND REWARDS

I

"THINK of the new knowledge!" said Cavor to Bedford, in Wells' First Men in the Moon, as the two discussed the reason for their flight into the unknown. In those words Cavor meant to sum up the spirit of the scientist, ready to risk and to sacrifice everything for the prospect of a new understanding of his universe. Upon a hope for this new knowledge, which might of itself have no immediate practical value, the case for the space flight must be based. Against it must be set the huge cost of the expedition. which, including the preliminary experiments, and the designing of the ship, may reach $100,000,000 (£20,000,000).

Although at first this appears to be a staggering cost for something so elusive as a peaceful conquest of the planets, the reader has already been reminded that it represented the equivalent of only two first-class battleships.

The cost of the expedition, then, is purely relative — to be compared only with the expected rewards. If peoples and Governments can be convinced that the stream of treasures poured into the building of the ship will return them a dividend of increased knowledge, if not of practical benefit, assurance can be given that the money will be available. What, then, can be offered by a space exploration in exchange for this emperor's ransom?

In the first place it would be only the most carping of critics who would insist on a shilling for shilling return for value expended. If that policy had been pursued we would have not a single astronomical observatory in the world to-day; many of the epoch-making discoveries in physics, chemistry, and even medicine would not have been made; and our colleges and universities would be obliged to empty from their halls a large number of their students.

Sir Arthur Keith, the eminent English scientist, stated recently that the important thing in science is a new discovery. The practical utilization of it has always come afterwards. Such should be the case with the discoveries that will follow the space flight. An interplanetary expedition, assuming that explorers can land on Mars and Venus, must be based alone on the prospect of those intangible rewards that arise from a deeper, wider, and truer knowledge of our physical universe. It must fall or stand on its ability to promote happiness indirectly by removing the veils of ignorance that make us strangers in a strange world.

The benefits from the space flight should flow from what man will see, observe, and learn. And until a world is willing to exchange two first-class battleships for the knowledge that the exploration of other worlds will provide interplanetary travel will not come. In fact, until then it should not come.

II

The benefits to science of an exploration of Mars, Venus, and possibly of the moon will for obvious reasons precede any others. The first explorations will undoubtedly be made by and for scientists, hungry for answers to problems that have long been before them. They will seek out missing facts with little regard for anything else; and upon their discoveries the fate of further voyages will depend.

The first and surest of these scientific rewards will go to the astronomer. The study of the planets and of their motions, the observations of the sun and of the fixed stars, would all be placed on a newer and more exact basis. Conclusions reached from our position, under the blanket of terrestrial atmosphere, could then be checked on the airless moon or nearly airless Mars. We need no longer suffer the disadvantages of the "fixed observer." We could if necessary carry out observations on three planets simultaneously, and verify beyond doubt conclusions reached only on earth.

Let us consider some of the riddles of the universe that our scientists are attempting to solve, questions that are raised by the incompleteness of our astronomical knowledge. Upon these riddles might be based the picture of what the conquest of space can do for astronomy.

First is the question of the extent of our solar system, brought to our attention recently by the discovery of the ninth planet, Pluto. This addition to our family, "by doubling the volume of the sun's domain, has raised more problems than it has helped to solve." And, according to Dr. Shapley, by this discovery "we are probably on the trail of a series of new planets."

The use of a telescope on Mars, 40,000,000 miles farther from the sun than the earth, should certainly aid tremendously in these discoveries. For not only will we have new places of observation unaffected by the wavering, flickering blanket of the terrestrial atmosphere (in space each shining world will stand out sharply and clearly), but the possible size of telescopes would be enormously increased.

To both the world of astronomy and the public at large, eager to know the domain of the little corner of the universe that is our solar system, the interplanetary voyage is recommended.

The origin of the solar system is the second of the problems, puzzling not only to the astronomer, but to the biologist, the chemist, and the philosopher. Upon this question rests much of our attitude toward the meaning and probable future of our earth. Were the solar planets born by the disintegration of a great spiral nebula rotating at tremendous speed? Or were our planets plucked from a greater prehistoric sun by the gravitational pull on its flaming surface of another sun that passed in its neighborhood aeons ago? Or did we originate apart from, but at the same time as, the sun — as sister and brother, instead of as parent and child? The answer obviously is brought closer by each increment of knowledge we can gain about the universe as a whole.

In its tremendous implications this question of how our solar system originated and what its future is likely to be is one that cannot but stir the curiosity of the most unimaginative. In the exploration of other planets, in the study of their geological formations, of their internal and external heat, their plant life and mineral deposits, determinations of the age of these worlds can be made. From these studies much will certainly be learned of the method and the order of the creation of the planets that will aid in our final understanding of the creation of the solar system .

The question of the extent of the earth's atmosphere is another of the unsolved riddles. Our blanket of air is, as we have learned, not only our protection against the violence of the cosmic forces, but also, in a sense, our prison. Its exact height is unknown, the constitution of its upper layers and their temperature are unknown. It is constantly shifting in position and extent in unknown ways; affected by the sun's internal convulsions, it produces great storms, which disrupt our means of communication, destroy property worth millions of pounds, and cause the loss of thousands of lives yearly.

If we are not to remain ignorant for ever of the great atmospheric forces that affect our lives we cannot remain hiding under our blanket of air hoping that nothing will happen. By venturing into and beyond the upper atmosphere we will probably be able to observe the causes of these convulsions of nature and to study their laws. By such means we can discover ways to anticipate and guard against them.

Penetration into the great spaces will permit a study of the sun itself — its spots, its surface, and internal changes as indicated by the spectroscope; and will afford a true picture of the effect of these changes on the atmosphere and on the earth's surface. If we could predict with certainty changes of temperature, snow and rainstorms, tornadoes and hurricanes, the saving to commerce and shipping alone would repay within a year the entire cost of the space-ship. . . .

III

A fourth riddle concerns the possibility of life on other worlds. It is one which intrigues the interest of not only the scientist, but of every one with a trace of imagination. We have seen that life may possibly exist on Mars and Venus. An interplanetary expedition will not only settle the question definitely, but will provide us also with an answer to the question of whether it might be possible for men to colonize and to live on them.

A fifth concerns, in a sense, the future of our own earth — for it is the question of the death of the sun upon which our lives depend. We have been hearing much lately about the tremendous streams of energy the sun is radiating into space. It has been doing that for millions of years. The sun we know today is only the remains of a younger and more robust body that once existed. How and why did the change come about, and what is likely to occur in the future?

The study of the sun from the vantage of other worlds, with immensely better conditions for observation than on earth, would bring the answer to this question much nearer. Complete studies of the radiation of the sun could be made and based on fact instead of on theory. We should no longer be confined to the bottom of a sea of air. Accurate and convincing studies of the life curve of the sun might give us some inkling of the destiny of the human race.

The location of the centre of our galactic system, a sixth riddle, and many others, including the questions of the ultimate extent of the cosmos; the nature, origin, and orbits of comets and meteors; the mysteries of space and the final summation of all these questions, the origin of the entire cosmos are the other problems whose solution will be hastened by the space flight. It would be naive to say with certainty that their solutions can be achieved by a study of other worlds and a study of the heavens from these worlds. But the scientist is well aware of the tremendous approach to truth that is made when studies that have been confined to a single locale are extended and diversified.

The gain is more than a proportionate gain in the number of observations. It is a gain not only of independent observations, but also of new avenues of thought stimulated by a new environment and a new point of view. If we are to assume that our astronomers will continue to attack these mysteries as diligently in the future as in the past, then it can be asserted that the space flight will save for them years of fruitless labour. The practical man who admits that astronomical knowledge is desirable or necessary cannot but be impressed by the financial if not the intellectual considerations inherent in this great saving of time, money, and energy.

We are now expending some millions of pounds in the construction of a 200-inch telescope to extend the range and power of our observation of the universe. What this great telescope is aimed to accomplish can be done also by observers on other worlds.

For, irrespective of whether Mars supports life of its own, it should prove a boon to our astronomers. It has two conditions so urgently desired by them. Its small gravitation — only one-third that of the earth — may permit the building of a telescope larger by far than anything possible on the earth. The tenuity of its atmosphere would permit observations far excelling in clarity, trueness, and range any that are possible with telescopes on earth.

On this question Professor Luyten states that,

> In the problem of stellar distances we are really entering one of the most vital points in our study of the universe. On our knowledge of distances rests our conception of the structure of the universe, and it may well be said that the determination of the distances of the stars constitutes the most urgently necessary task, and at the same time the most difficult task in all practical astronomy.[1]

The distances of the stars are measured in two ways — the triangulation method — using as large a base-line as possible (at present the diameter of the earth's orbit), and spectroscopic observations. An observer on Mars would have a base-line of 285,000,000 miles, as compared with 185,000,000 miles on earth, which would extend by 50 per cent, the study of stellar distances by this method. He would also have, by the elimination of the atmosphere, which absorbs some of the spectral lines, unexampled opportunity for the determination of distances by spectroscopy. Questions that are arousing the world of astronomy, such as the bending of light rays near the sun, the earth-sun distance, and the stellar motions, could all be aided to a solution by an observatory on Mars or by observations in space.

These rewards to the astronomer include only suggestions of how a more accurate, more far-reaching, and more diversified knowledge would assist in the solution of the problems that so intimately concern him. It opens up great new fields of effort and opportunity, such as he might never have dreamed possible. The ultimate construction of a great astronomical observatory on Mars is certainly a reward the space flight can offer.

1 Willem J. Luyten, The Pageant of the Stars.

IV

The biologist too is concerned with cosmic verities. He starts his investigations with microscopic forms of life instead of gigantic flaming suns, and for that reason the experiences he might wish to meet with on other worlds will differ from those desired by the astronomer.

The biologist wishes for a wide variety of life forms. The astronomer is not interested in them. According to Dr. Shapley, life is but one of the many crustal phenomena at the surface of a planet, and the planet, to the student of sidereal affairs, is a fragment of secondary importance. Animal behavior is a trivial matter compared with the behavior of elementary gases. Human laws are transient, weak, arbitrary, and absurd when contrasted with the impressive laws of physical science.... For an understanding of the universe the radiations of a comet's tail are more important than living organisms.

What the astronomer desires is a perfect opportunity for the observation of distant stars, gigantic nebulae, and star clusters. Since the earth's atmosphere hinders these observations he would wish for an observatory on a world with no air blanket above it. His desires, then, conflict with those of the scientists who study life forms, since a world without air could support no life such as we know.

Whereas Mars is primarily for the astronomer, to the biologist and paleontologist, the anthropologist, the chemist, and the student of the medical sciences, Venus may be a veritable laboratory Paradise.

The biologist wishes, above all, to know how life began — how inanimate chemicals became transformed into breathing, digesting, thinking, feeling creatures, and what the factors are that govern their existence. Upon the solution of that problem will rest much of our belief as to what human beings are and what is their place in the cosmos.

Is life, as Sir James Jeans suggests, merely "the result of a disease that attacks matter in its old age?"[1] Is it only an accident happening on one little planet out of the billions that must exist throughout the universe? Or is it the natural flowering of certain physical and chemical forces, whenever and wherever conditions are favorable?

For an answer to the question of how and why we came to be, what could be better than the stimulation provided biologists by a study of entirely new forms

1 Sir James Jeans, The Universe Around Us.

of life, produced under radically differing physical conditions?

Says Jeans on this question:

> What we would like to know is whether [life] originated as the result of still another amazing accident or succession of coincidences, or whether it is the normal event for inanimate matter to produce life in due course when the physical environment is suitable. We look to the biologist for the answer, which so far he has been unable to produce. The astronomer might be able to give a partial answer if we could find evidence of life on Mars or some other planet, for we should then at least know that life has occurred more than once in the history of the universe. . . .

Sir James puts the burden on the astronomer because he undoubtedly looks to ever stronger telescopes to bring our sister planets closer for observation. But what telescope can bring them close enough for studies of plant and animal life even for the physical detection of this life? The answers to the questions Jeans propounds must rest upon an actual expedition to other worlds, where the biologist can not only see, but study, compare, and experiment, and arrive at conclusions based on pragmatic knowledge.

Much of our philosophic thinking will depend upon the result. If we find by an exploration of Mars and Venus that life is indeed localized upon the earth, then belief in special creation — of a newer, higher, and more cosmic kind — might find scientific validity. In the minds of those whose intellectual horizons extend beyond the everyday problems of their country and age the puzzle of life's beginning must indeed provide a perpetual source of speculation to which an answer is deeply desired.

Another question of similar implications is that of the life and growth of plants upon which our lives depend.

We look to our plants not only for food, but also for the replenishment of the oxygen of the air used in our bodily chemistry. Plants absorb the carbon dioxide in the air and release oxygen. How do they do this? What conditions are necessary for this action? What would stimulate it or stop it altogether?

Further, how does the plant, by photosynthesis, transform the carbon dioxide into starch and cellulose? The internal life of the plant and its use of the sun's energy in the process of photosynthesis is still almost a complete mystery to us, and we can only hope that nothing will occur to either plant or sun to destroy the delicate combination on which our lives depend.

The study of plant life, its growth and life cycles on other worlds, would accomplish one vitally important thing. It would isolate the purely local factors that operate on the earth, and make us aware of the more universal forces acting on plant life. We may find that plants elsewhere are stimulated by certain rays of the sun that do not reach the earth — or that they may be killed by these rays. Certainly the effect of the cosmic rays not only on plants, but on all life, would be determined once and for all.

Some physiologists have concluded recently that it may be the cosmic rays that cause death by old age, by wearing out the tissues and leading to the decay of the organs. If we were more adequately protected from cosmic rays, it has been suggested, we might be able considerably to extend the normal span of our lives.

What wonderful opportunities for new knowledge, so vital not only to our intellectual conceptions, but to our everyday lives, is opened by the space flight! A new control over life and happiness is indicated by the worlds of fact that are opened before the scientist.

V

To the chemist and physicist is offered the study of matter and its nature and possibilities not only on other worlds, but in space itself. Says J. H. Fabre:

> Part of the sun's rays reach us in an enfeebled state apparently because they are extinguished on the way. This is caused by the sun's atmosphere and our own. There is reason to believe that the atmosphere of the sun includes metallic vapors corresponding to nothing known on earth, for many dark lines of the solar spectrum do not coincide with any that are produced by terrestrial substances.[1]

Our own list of the elements that are supposed to comprise all matter is, perhaps, not complete. There may be others in the sun whose nature, qualities, value, and effect may be still unknown. Our physical science is perhaps only an approximation based on observation limited to our exclusively local position. We are forced, by being tied to earth, to use all our instruments of observation under bad conditions. The physicist stationed on Mars or the moon would be enabled to study unknown elements directly, with no terrestrial atmosphere to "enfeeble the rays" and to bar out others. What these studies might mean to our sciences of physics and metallurgy, and perhaps also to manufacturing, we do not know and cannot guess. They may establish more solidly our present theories regarding

1 J.H. Fabre The Heavens (Lippincott).

matter, or may radically alter them. They might put us on the track of substances of great value.

The chemist on a Martian or Venusian exploration could examine the elements and the combinations of elements of which the planets are composed. He might find stores of materials which are scarce and valuable on earth, such as radium. This might be particularly true on Venus, where the apparent youth of the planet has not given time for so much of the precious stuff to decompose. On Mars, an older world, he might find that time has wrought curious new combinations of matter, with results of vital importance to us on earth. Certainly our own knowledge of nature is pitifully small. It is inconceivable that discoveries on other planets would not add greatly to it.

To the physicist the increase in knowledge would open a treasure-house of facts on the ultimate nature of matter and its force relationships. Studies of cosmic rays and other radiations and extended observations on the destruction and rebuilding of matter in far-off suns would offer new evidence to verify or to destroy atomic theories. To medicine there may be offered new worlds of plant and animal life, and opportunities to study the effect of new physical conditions on organic life. On another planet what strange new manifestations of bacteria and disease, offering him freshness of view-point and new standards for comparison with his present knowledge, might be discovered by the physician! Such learning, in a practical field such as medicine, must ultimately be translated into prevention and cure of disease, extension of life and promotion of health on the earth.

Scientific rewards that might be made possible through the space flight could be extended indefinitely, for they run through the whole field of our knowledge of nature.

Because he is so suggestive and stimulating, Sir James Jeans may be quoted again to conclude this phase of the interplanetary question.

> " The three centuries," he says, "that have elapsed since Giordano Bruno suffered martyrdom for believing in the plurality of worlds, have changed our conception of the universe almost beyond description, but they have not brought us appreciably nearer to understanding the relation of life to the universe. We can still only guess at the meaning of this life which appears so rare. Is it the final climax toward which the whole creation moves, for which millions of years of transformation of matter in uninhabited stars and nebulae and the waste of radiation in desert places have been only an incredibly extravagant preparation? Or is it the mere accidental and possibly

unimportant by-product of natural processes which have some more stupendous aim in view?..."

These questions touch the heart of all knowledge and all philosophical thinking. The past centuries of man's life have been an adaptation of the tools of nature to give him safety and comfort in his daily life, to protect him from the elements, and to establish means for communication. Man's vision now extends not only outward to further conquests ending in better living, but also inward to an examination of what he is. Should these aims be realized future centuries might indeed be Golden Ages. An incredibly vast new world of life and thought is opened up by the prospect of the space flight and the exploration of other planets.

CHAPTER XVII

A GLIMPSE OF THE FUTURE

I

LOOKING ahead to the year 1950, one may confidently envision long-distance rocket-planes engaged in regular flights over the far places of the earth.

The technical problems of controlling rocket fuels should be successfully overcome within five years. An additional five years should certainly see the completion of simple rocket projectiles capable of ascending fifty or more miles into the air. The following ten years allow sufficient time for the design and building of rocket-planes and for the training of pilots and navigators in their operation.

Industrial organizations are no doubt ready even now to act upon any evidence that the rocket has proved itself as a definite instrument of transportation.

Funds and facilities will not be wanting for men able to unite the elements of the rocket into a practical instrument.

The inventor will turn over to the engineer a plan for a rocket vehicle. The engineer will design it, and capital will be found to build and to operate it. When these necessary steps in a new invention have been passed giant rocket-planes will roar across the sky with speeds so swift that present-day aeroplanes will seem sluggish in comparison.

Rocket-planes, carrying first mail and later passengers, will make of the earth but a small nation. World-centers of population will be linked with rocket lines, for oceans, mountain ranges, jungles, and storms will not impede the rocket's flight.

Great rocket-ports, of the passenger capacity and complexity of metropolitan railway termini, will be located outside populous cities.

Such a terminus might be found on Long Island, New York, in the year 1950. Let us imagine the scene on that great tract of land. Metallic streamlined rockets drone down from the skies. From the ends of the earth men and goods are being brought to New York with dizzying swiftness.

From gleaming metal-sheathed hangars the rocket planes are wheeled out into their places to receive passengers and freight. Side by side rest these great gleaming birds. LONDON-NEW YORK is emblazoned on one. On another PARIS-NEW YORK and on a third BERLIN-NEW YORK.

It is nine o'clock in the morning. Aeroplanes are already landing in a steady stream at the rocket-port to discharge the passengers for the European expresses. Coming from the smaller cities, the passengers alight at one end of the field and are carried a quarter of a mile by bus to the transatlantic giants.

Spanning the ocean in hardly more than an hour, the rocket-plane makes of a trip between hemispheres a journey no more considerable than a short train ride in our own day.

All of the planes carry mail, much of it having been posted in New York the evening before. Within half a day letters will be delivered at London, Paris, or Berlin addresses. "Home thoughts from abroad" need no longer be the sentiments of one who is exiled, for the home is but a few hours away.

Appearing far in the northern heavens, a 'plane swiftly grows in size. In a few minutes it lands a group of men and women from a midsummer weekend in Alaska. Another from the east brings several dozen university students, who in two days since leaving New York have visited Southern Italy and the antiquities of Northern Africa.

II

Almost invisible, a 'plane traces a half-circle of flame from south to north across the sky. The timetables show that it is the Ottawa-Washington express returning Canadian statesmen from an important overnight conference at Washington.

The western edge of the field receives at this hour its rocket-omnibuses. Travellers to New York from Massachusetts, Delaware, and Pennsylvania are being landed. They transfer to autogyro-planes, and will be in the city in ten minutes, having completed a four hundred-mile journey in three-quarters of an hour.

Equipped with the Goddard rocket-propeller systems, these ‘planes offer no danger from their exhausts to those on earth. The ‘planes are lifted from the ground and landed by propellers turned by the gases of the fuel combustion. Only when the ‘planes are well into the air are the propellers withdrawn into the nose and the gases sent roaring from the exhausts.

In cabins fitted with comfortable upholstered seats, free from the shock and vibration of the rocket motors, travellers across the continents have a restful journey. They are acquiring a new concept of distance. Space to them is but a function of time, and journeys are measured in hours and not in days.

Rocket-plane passengers will learn some of the peculiarities inherent in our systems of time measurement. The German bound for New York will leave Berlin at noon, and as the ‘plane speeds through the higher rarefied air he will see the sun sink in the east. Lower and lower it will descend, less intense will become its light and heat. In two hours New York will be reached and our traveller will find that it is but eight A.M.![1]

He will have travelled four thousand miles, and in addition have added four hours to his day. But this gain must be returned when he again travels eastward. Leaving New York at noon, perhaps on the same day (after he has transacted his business), he will see that the sun rushes westward in the skies with alarming swiftness. An hour after leaving New York the sun will be already half-way toward the horizon. Then swiftly will come the sunset, and with terrible suddenness the night will fall, as he nears Berlin. Noon and night within two hours for it will be eight P.M. when the rocket lands at the Berlin port!

A day in 1950 may see the area about Times Square, New York, flooded with thousands of people watching breathlessly the moving illuminated words of the New York Times electric news-recorder. A rocket flight, to race the sun round the earth, is in progress. All business and all traffic in this congested neighborhood will have ceased as hour after hour the reports of the dash of the flyers round the globe are being flashed.

1 Time at Berlin is six hours earlier than that in New York, when it is noon in Berlin it is but six a.m. in New York.

Leaving New York at noon, the flyers will reach San Francisco after an hour's flight — when it is only ten A.M. With the journey hardly begun, they will already have left the sun far behind. Transmitting their progress by wireless, the flyers will tell the world how the sun sinks swiftly in the east.

The electric sign will inform the hypnotized crowds that the flyers have transferred to a second rocket-plane at San Francisco, and without delay they have roared across the Golden Gate, swiftly crossing the Pacific on the "long" two and a half hours' flight to Tokyo. The sign will tell how the flyers found that, like a dream in which everything moves backward, the moon faded into morning as the shores of Asia turned toward them. For when Tokyo is sighted it will be but eight-thirty A.M. local time.

As the flyers speed across Asia toward Moscow in a third 'plane they are still turning time backward in its flight. The grey light of dawn at the thirty mile altitude is giving way to night. In darkness lit only by the flame of the exhausts the rocket will fly high above the earth until, when Moscow is reached, it will be but One A.M.

Destroying continuously the bulwarks of time, the flyers will rush westward, spanning Europe in a leap and watching, many miles below, the broad Atlantic as it seems to roll eastward. The night becomes less intense. Below them the sea reflects the twilight of an approaching evening. But as they move on its shades become imperceptibly lighter and lighter.

Seven hours after leaving New York, at seven P.M. New York time, the flyers will have reached their goal. While the straggling sun is still wearily crossing the Pacific crowds at the Long Island rocket port will be greeting the conquerors of space and time.

III

Apart from the tremendous effect that rocket. planes may have upon international transportation after 1950, they are certain to make vital changes within a geographically large nation, such as the United States. With commutation to a great city becoming possible from an area three to five hundred miles in radius, a spreading out of population into less-populated districts is sure to come. New communities hundreds of miles from the city will arise, with the travellers owning possibly a rocket-omnibus. These little centers, built as a revolt against the crowding of the city, will act as spurs to the rebuilding of the great cities upon a more spacious plan.

Some city-planners have seen the city of the future as consisting of a restricted number of great buildings spaced far enough apart to permit beautiful gardens among them. An incentive is needed to make of this vision a practical programme.

If the aeroplane, and later the rocket-plane, can provide workers with an opportunity to live far from a city's noise and squalor and narrow confinement, they will have made possible the building of newer and more beautiful cities.

It is not expected that the railway and the steamship, though they will seem unbearably slow to the man of 1950, will vanish. They will still serve as roomy means for passenger transportation and carriers for the heavier and more bulky freight.

It is not expected, furthermore, that the "speed craze" of the present era will grow in intensity with the increases of possible speeds. There will be a change instead in men's conceptions of distances. As Michael Ardan said in Verne's From the Earth to the Moon: "Distance is but a relative expression, and must end by being reduced to zero."

The man of 1950 will not have a frantic sense of being whirled dizzily through space in the rocket. Rather his journey will be made with the cool realization that 4000 miles is equivalent to a trip of a couple of hours. The distinction is important. In the rocket-plane he will have no clear sense of the speed of his vehicle. The earth, contracted beneath him, will roll backward to his onward flight, and cause him to realize forcibly that ours is but a little globe.

The growing conquest over terrestrial space by the man of 1950 will, as was suggested at the beginning of this book, provide him with the mental vision necessary to conquer the interplanetary spaces. Although at the present time the interplanetary journey is within the bounds of physical possibility, our experience with cosmic speeds and distances is not equal to the task of guiding a space-craft on its perilous journey.

Perhaps by 1950 the brilliant project of Hermann Oberth for a station in space may be interesting the world-wide attention of engineers. The proposal of Oberth to build an artificial satellite of the earth, if found feasible, may provide the mental and physical stepping-stone from our conquest of the earth to that of the solar system.

The station in space would, according to Oberth's plan, be a giant rocket shot into the air, which, attaining a speed of five miles a second, would circle the earth at an altitude of five hundred miles. This speed, sufficient to keep the rocket from falling back to earth, would cause it to circle the earth for ever without further power.

Making a complete revolution round the earth in one hour and three-quarters, the station in space would provide a perpetual symbol of man's defiance of gravitation. Men stationed on the satellite, provided with the necessities of life and the best instruments of science that the ship can carry, could supply the earth continuously with the most exact information about nascent storms, cyclones, and movements of ice formations. The extension of the work of the Goddard meteorological rocket might then be made on a grand scale, with observers continuously at work on studies of the sun, of our sister planets, of the Kennelly-Heaviside Layer.

As Oberth has stated, these satellites (for a number of them might be built to circle the earth in different courses) would become "Argus eyes" of the earth. They might become our contact with the rest of the universe.

The information gathered by observers in the satellite might be transferred to earth by light flashes, or by wireless or infra-red beams. Changes in the personnel and provision of supplies and equipment would be sent from earth by smaller rockets. The supply-ships would rise from the earth, fly parallel to the station, and be gradually drawn toward her. Physical contact might be made by means of magnetized plates, and then by air-locks between the two ships.

The projection of a station in space is admittedly a gigantic task, and might prove infeasible. It may be discovered that the attraction of the moon and the sun, at various points in its orbit, might throw it off its course and either pull it farther and farther into space, or send it crashing back to the earth's surface. Danger from meteors and cosmic rays, the intense heat of the sun, the problems of keeping the station air-tight and supplying its crew with the necessities of life, might be too tremendous for the twentieth century to meet successfully.

Yet these very difficulties must be faced in the building of a ship to make the journey to the moon or the planets. It may be solved first in the station in space.

IV

The use of rocket-planes for tremendously swift transportation will undoubtedly suggest to military men their adoption for warfare. With great speeds so easily attainable the rocket-plane may vitally supplement the aeroplane in making attacks upon enemy troops and cities.

Two hours after a war has been declared a combatant nation might without warning find its cities bombarded by a fleet of rocket-planes. No provision for defense could be made in time to prevent this fleet from wiping out great areas of population. The rockets, flaming out meteor-like in the night, would loose their tons of destruction, and, circling the skies, vanish to return swiftly to their starting place.

Because the rocket-planes can fly at any level no effective defense against their coming can be made unless the skies are swarming with protecting planes. Because their speed will be so great the chance of hitting or destroying rocket-planes will be extremely small.

These facts are presented with no relation to their moral significance. The engine exists, and will undoubtedly be used in a future war, perhaps with calamitous effects. Short of the cessation of warfare, there is no apparent way of barring its use.

The rocket of 1950 may then be an agent of great good and of equal harm. Whether men twenty years from to-day will wish that it had never been discovered we have no present means of knowing. By the creations of our hands and our brains we have accomplished our difficult climb from savagery to our present civilization. By such a creation as the rocket, aided by our own wilfulness, we may fall just as swiftly back to savagery.

On the other hand, we know that the rocket opens new worlds immeasurably vast in their promise. Barring accident, or a catastrophic war, the man of the future should see, at least in part, the realization of that promise.

THE LANDING OF THE ROCKET SHIP ON THE PLANET MARS, AND THE GREETING OF THE MARTIANS AS CONCEIVED BY JEX. THE POSSIBILITY OF LIFE ON THE RED PLANET IS AT LEAST A FASCINATING THOUGHT.

BY ROCKET TO THE PLANETS

BY DAVID A. LASSER
PRESIDENT, AMERICAN INTERPLANETARY SOCIETY

Are we on the threshold ready to reach out into space for new knowledge? Is man soon to go outside the atmosphere of Earth and travel by rocket ship at great speed, even to other celestial bodies? Mr. Lasser, who is President of the American Interplanetary Society, thinks so and has written this article about what he thinks. Undoubtedly it contains physical conclusions susceptible of debate, but this story may, at least, stir the imagination as to what Science has before it. Incidentally, our artist, Mr. Jex, wishes to disclaim any intimate knowledge of Mars and interplanetary travel. In fact, he has never made such a trip and says his illustrations are pure figment.

TO THE imaginative mind, viewing the star-filled heavens and the luminous glow of the planets, there is present continually a world of mystery, and the suggestion of tremendous adventure beyond our earth. Since the day man recognized the planets as worlds like our earth, two questions have filled his mind: how can I get there and what will I find?

In myth, in story, in scientific speculation, literature from the time of the Greeks has been filled with commentaries on the nature of our sister planets and of the great interplanetary space. Investigators searched out with telescopes the mysteries of the heavens, building up laboriously the impressive science of astronomy. By their side were men of letters, with unleashed imaginations, peopling the heavens and the planets with the creatures of their fancies.

In all this investigation and romancing, it was not questioned that some day the planetary mysteries would be solved. Man would travel to the moon on chariots propelled by "swans of the East Indies", as Bishop Godwin, writing in the Seventeenth Century, put it, or in a balloon, as Poe pictured it in his "Adventure of One Hans Pfaal", or, again, in the cannon projectile of Jules Verne. He would then discover for himself the forms taken by Nature's creations.

These dreams of scientists and men of letters alike, until the dawn of the present century, were not burdened by the grim difficulties of the interplanetary journey. It was not until the nature of interplanetary space was determined that the fanciful means of interplanetary travel were abandoned and a real science of spatial navigation erected. Similarly, it was not until high-powered telescopes, and the spectroscope, had been trained on the moon and our sister planets that any speculations about their inhabitants yielded to analysis of what life they really could support.

Fancy, then, has been displaced by science, and through science man today has in the rocket ship a vehicle that in theory at least can lift him away from the earth across airless and heatless interplanetary space to the moon a quarter of a million miles away, or to the distant planets, Mars and Venus. And in a half dozen nations scientists of the highest repute are laboring on the giant that slumbers in the rocket to make the theory a reality. When their work has been completed, and an interplanetary ship is built, the experiences that man will enjoy on his journeys into the great spaces and upon our sister worlds will no doubt be more fanciful than our wildest dreams!

LOOKING BACK UPON THE EARTH FROM THE ROCKET
Our globe as it might appear to the occupants of the interplanetary ship en route to other spheres.

To appreciate the nature and significance of an interplanetary journey, it is necessary first to understand the craft which is to transport us. The rocket in simplest terms is a form of recoil instrument. A charge of explosive is burned in a combustion chamber, and the resulting gases exert a tremendous pressure upon the chamber walls pushing the rocket forward at the same time that they are exhausted to the outside. The action is similar to the recoil of a rifle — the explosion of the powder not only sends the bullet on its way but also kicks the gun emphatically against the rifleman's shoulder.

By its ability to accelerate slowly in order to arrive at the speed of seven miles per second, necessary to escape fully from the earth, and because it operates in the vacuum of interplanetary space, the rocket is peculiarly fitted for the great task imposed upon it. It can make the fullest use of the tremendous powers latent in the explosion of hydrogen and oxygen, as well as other fuels, and combat successfully the relentless pull of the earth upon its children who hope to escape her.

As yet no rocket fuels have proved in test that the technical problems have been solved. Nor have rockets in flight gone high enough to yield revolutionary results. But month by month during the past two years progress has been evident. From flights of a few hundred feet, rockets have ascended six thousand feet, and plans are now in progress in Europe and America to send them up from five to two hundred miles. A recent grant to Dr. Robert H. Goddard, an eminent American physicist, of $100,000 by the late Daniel Guggenheim is allowing Goddard to carry out with freedom his ambitious rocket experiments.

Also working in America is the American Interplanetary Society, lending organized support to the interplanetary project and planning experiments of its own. Abroad, the powerful German Interplanetary Society has made that nation rocket conscious, and at its rocket flying field in Berlin its engineers are making headway against the numerous technical difficulties. In France, under the leadership of Robert Esnault-Pelterie, a distinguished aviation engineer; and in Russia and Austria, other groups are attacking vigorously the barriers that block our way to the heavens. The dream of the interplanetary flight that was once a weapon for satirists, a butt for ridicule, a dream of theorists, is now on the way to becoming an experience possible for the race!

To convey to a reader a true picture of an interplanetary journey would require a great deal of space, because there is little in our own experience to equal it. Then, too, since we have never travelled into interplanetary space, our picturing of the sensations and experiences must be drawn heavily from conjecture.

We do know that the start of the journey, whether to the moon, or Mars or Venus, will probably be made at night, and at a rigidly predetermined hour. The flight will be based on a mathematically plotted curve through the heavens. There we will have no mapped road, and only the heavenly bodies as signposts. We must use their positions in the heavens to guide us to our destination on a path as exacting as science can make it.

At the start we will be lying in a metal-walled cabin of the rocket ship, in comfortable bunks, because the acceleration at the start will be so great that an unbearable pressure would force us to our knees, were we standing. Lying down, we can breathe easily until at a signal from the navigator a great roar sounds behind us, and simultaneously the hand of an invisible giant presses us to our cots. For ten minutes, this pressure, growing more and more unbearable each moment, continues. It is the pressure of our acceleration through the atmosphere of the earth, and it is as if a load of twice our own weight were resting on our bodies.

FREE OF EARTH
The rocket sails into space with its course charted to the moon, Mars or Venus.

But at the moment when it seems that the pressure could not be borne longer, it gradually begins to relax. The weight is removed by the invisible giant, and we begin to breathe more and more freely. The roar from the rocket exhausts has somehow ceased and we learn that the power has been shut off. A speed of more than seven miles a second having been attained, we are forever free of the earth.

We look out of a window at the side of the cot and see with amazement that our earth, fifteen hundred miles below, has shrunk appreciably in size. We are looking upon its night side, and slowly, ever so slowly, it turns its darkened oceans and continents toward us.

Then another sensation becomes unpleasantly evident. The pressure resting upon us has wholly disappeared. But then it is as if the cot beneath us were gradually withdrawn, or as if we were lifted gradually from it. A feeling of a lack of support beneath us becomes painfully clear, and for an anguished moment one feels as if he were falling through space!

But the comfortable walls of the cabin are still about us; the sensation of falling, we learn, is that state of "weightlessness", just an accompaniment of flying through space without power. There is not a sound, no sense of movement, no single indication that we are fleeing from the earth at two hundred times the speed of an express train.

When we get up, it is with care, for since we weigh nothing now, a single false step would send us shooting painfully to the ceiling of the room. And as we discover later, our whole lives must be reorganized to these strange conditions of space. We cannot eat, drink or bathe as on earth. Water, being weightless, will not pour, nor will it remain comfortable in the bottom of a tub or basin. All free liquids will float about a room disconcertingly in small globules; and to drink, one must squeeze his liquids from a soft, flexible flask. And to breathe, in this sealed vessel, we have a ventilating system that automatically sends through every room a stream of air, of the pressure and constituency we are accustomed to on earth. Impurities are changed chemically and the air revitalized for further use.

Hour after hour the ship will speed on silently, because its trip to the moon will consume possibly a day and a half. For a journey to Mars or Venus thirty to ninety days may be necessary. The trip in either case will need all of one's faith in the trustworthiness of the ship and the ability of the crew to take him safely to the longed for destination. One will not travel directly toward Mars or Venus, but rather toward a point in the heavens where these bodies will be when we arrive. While we are moving these worlds are moving too, and our course has been plotted to effect a meeting with cruel accuracy. No room is left for chance or dallying in space. Disaster by a collision, or being lost in space with no fuel is the alternative to tampering with the iron laws of spatial navigation.

But the hours, or days, of the journey will pass. Aflame in the sky, the intolerably blazing sun will throw hundreds of miles into space its gigantic prominences, while ringing the orb will be its incomparably beautiful corona, seen on earth only at times of an eclipse. The sky, even behind the sun, will be a deep, fathomless black, while the stars, billions of them, now, will shine out with a brilliant, hard, steady light.

A day will come when we will feel ourselves to be flotsam in an eternity of emptiness. Time, motion, space will all have ceased to have meaning.

As day after day passes in the changeless sky, doubts will arise as to whether Mars or Venus will be at the appointed meeting place at all. The earth will have shrunk to a ball, then to a disc, and finally to only a glowing point of light in the heavens. The emptiness will seem infinite. We will take observation after observation on the stars; check and recheck the speed and course, and wait feverishly for the approach of the new world. Doubt will change to fear and then to terror as the weeks bring only increasing loneliness and monotony. A hundred times we will believe ourselves lost, and like the first navigator of the Atlantic we will be assailed by the panic-stricken crew to turn back before we perish miserably in the infinite space.

We will have to fight these things, against a certain doubt of one's own sanity as the terrible stillness of space envelopes us. But if faith triumphs, we will at last find its justification in the growing disc of Mars or Venus in the skies. We will see it day by day, as its path gradually converges with that of the ship. It will grow from a small to a large disc, then into a gradually expanding ball. From an unblinking star, it will become a round, fair young world. And as we adjust at last our course to circle the planet or land on its surface, we will find that we have acquired from this experience the steel and strength of a cosmic faith which we will hold forever.

To ask what we might find upon the moon, Mars or Venus to compensate for the labor, expense and torture involved is to expose our ignorance and our unconscious desires.

We know that we hope to find, most of all, a form of reasoning being not too unlike ourselves, with whom we may establish some communication. The thought of possible life upon the planets arouses at once the picture of something resembling human beings. But, sadly enough, we ask too much to expect Nature to duplicate upon another world, of totally different conditions, the product of millennia of earthly evolution. That we shall find life at all, animal or vegetable, is enough to excite the imagination and make the journey worthwhile.

We know quite definitely that the visible side of the moon is airless, waterless and gives little promise of any life upon its surface. But what may exist on the side that has been perennially hidden from us, or what may exist beneath its surface, we do not know. We do know that horrible extremes of temperature prevail upon the moon, and even without the airless condition of its surface it could not support life of any kind with which we are familiar.

But on Mars and Venus, the hope of finding life forms, even in an advanced stage, is quite good. We know that Venus has an atmosphere, though its constitution close to the surface is unknown. Its surface temperature is within a livable range, and there is undoubtedly water on its surface in which life, such as we know it, might germinate. But the form that such life may take may be beyond our wildest imaginings.

It has been suggested that Venus is a younger world than the earth, that it is in the condition the earth was ten to twenty millions of years ago. If that is true, it is possible that it possesses a great variety of primitive life forms, comparable in a remote way to those that once graced the earth. Some of these may have developed under the peculiar Venusian conditions rudimentary states of intelligence. But certainly, if Venus is the semitropical world that its temperature and thick cloud layers indicate, it will be a veritable paradise of wonders for the naturalist as well as the anthropologist and experts of other natural sciences.

On Mars, too, physical conditions permit the existence of life, though such life would probably be of a specialized character. For, with its extremely thin atmosphere, — containing only a small fraction of the oxygen and water vapor of the earth's blanket, and its temperature range from 72 above zero to 40 degrees below (this range often being covered between noon and midnight of one day) only a life form of the greatest hardihood could exist. It must be able to live on a curtailed ration of air, such as is found upon the tops of our own mountains, and be able to adjust itself quickly to the extremes of daily temperatures. If a life form such as man once existed on Mars, he would have either burrowed himself into the planet or fought a desperate race on the surface between the evolution of his being and death, as Mars lost its air and water. It is possible that a race resembling man might once have ruled the planet but failed in the race with evolution and gave way to less intelligent but more adaptable species. And if man wished to explore the red planet he must encase himself in a space suit (a modified diving suit) and carry with him containers of oxygen for breathing, to supply the deficiencies of the rare Martian air.

The meaning of the "Martian canals" discovered by Schiaparelli and Lowell and so ardently disputed by astronomers, will also be settled when an expedition lands on the red planet. We will know whether they are irrigation projects of a gigantic size executed by an intelligent race to keep alive, by herculean efforts, upon a dying world; or whether they are telescopic illusions or meaningless markings across the planet. Lowell died with the belief that Mars, being an older world than the earth, had once supported a race of intelligent beings who built the vast Martian canals. A settlement of this question should be itself sufficient reason for the journey.

The speculations that might arise on this subject are endless and it is possible that the final answer may escape our most shrewd guesses. But man of the twentieth century does not have within his grasp a means to the answer. He stands upon the threshold of adventures far greater and more thrilling than has ever been offered to the race. When the fulfillment will come, no one knows. But the instrument — the rocket — is ours, and with it the conquest of space should be possible.

Editor's Note

During the course of compiling this volume I was assisted greatly by Mrs Mimi Lasser as well as Mr Michael Ciancone. Together it was decided that a modest selection of Mr Lasser's papers would be included for historical perspective. The following pages include:

Article written for the Herald Tribune July 13th 1930 about Guggenheim grant to Goddard.

Proposed Constitution of the International Interplanetary Commission written in 1931.

Annual Report to the American Interplanetary Society from April 13th 1931.

The Rocket and the Next War written between the two world wars.

Letter from Edward Pendray to David Lasser regarding "The Conquest of Space".

Correspondence between David Lasser and Sir Arthur C. Clarke regarding AIAA.

Robert Godwin (Editor)

Guggenheim Financial Aid for Goddard Rocket Study Gives New Hope to Man's Efforts to Conquer Gravity.

A practical use of the long distance rocket will be made to make records automatically of atmospheric information that cannot be obtained any other way. Precision instruments that will register data of all kinds will be placed in the rocket compartment, instead of passengers. A huge parachute will bring the rocket back to earth safely with its cargo of recorded information.

With Vehicle Perfected - Science Hopes to Plumb Mystery of Outer Space.

Development of Plane and Radio Hinge on Factors Now Out of Mind's Reach.

By
David Lasser

President American Interplanetary Society

Since the dawn of science man has dreamed of escaping from the gravitational grip that holds him to the earth and penetrating the atmosphere to explore the worlds about him. And although that dream still is far from realization, signs are not lacking that progress is being made, at least toward the exploration of our terrestrial neighborhood. The recent announcement of the donation of a fund by Daniel Guggenheim for the study of the rocket by Dr. Robert H. Goddard, of Clark University, foreshadows the construction of a means of discovering the mysteries of our atmosphere and what lies beyond it.

Planes Futile at Great Heights

By means of balloons, the airplane and the dirigible, flights have been made away from the earth to make meteorological tests and discover what conditions exist in the upper strata of the atmosphere. But these means of ascension have in themselves their own inadequacy. Because of the increasing rarity of the air as they went higher, balloons and dirigibles lost their lifting power, while the airplane found no air on which its propellers could exert traction and on which its wings could be supported.

Almost simultaneously with the development of the airplane came research work on rockets. As far back as 1915 Dr. Goddard, working in a Princeton laboratory, was beginning to plumb the possibilities of the rocket, a relatively ancient means of shooting projectiles in the air.

Because the rocket obtains its propulsion from the recoil of gases expelled from its rear and does not depend on air for its activity, Dr. Goddard saw in the rocket the ideal means of ascension for high altitude work. In fact, he has stated, the rocket obtains its greatest efficiency in a vacuum. And the space beyond our atmosphere is supposed to be in vacuo.

Fifteen years of research and development of the rocket by Dr. Goddard have now been rewarded by what appears to be a liberal grant of funds and the support and assistance of a powerful and influential board of scientists. The way all being cleared therefore, for an escape by means of the rocket from the chains of gravitation and the blanket of atmosphere that covers us. A rocket carrying all kinds of delicate instruments will then be enabled to learn what lies fifty, 100 or 500 miles above us.

Many Questions Wait Solution

The questions that the rocket can help to settle are many. There is the mooted question of what outer space consists of — a matter over which the world's greatest scientific minds have been in disagreement. Is outer space — that which lies beyond our atmosphere and between us and other worlds — a complete vacuum, devoid of all matter and all heat? Or is it, as Eddington suggests filled with matter in a state of the utmost tenuity? The question is of more than academic importance for upon its answer depends the substantiation of some of the newer theories regarding space and matter.

How far does this matter extend and what is the nature of its upper layers? Does the blanket of air which protects us from the terrible heat of the sun and yet enables us to retain what we get, extend 100, 500 or 1,000 miles above us? What are the gases which circulate on its outer fringes?

These questions, too, are pertinent to man's knowledge of the influences that determine weather conditions and that influence more or less indirectly the success of airplane travel.

Astronomers and physicists attempting to see through the barrier of this layer of air and examine our heavenly neighbors, near and far, can gain new hope, for the rocket carrying instruments of observation will be enabled to view with mechanical eyes the undiminished brilliance of the sun, with its immense prominences and enigmatic corona, as well as the planets and distant stars seen by their naked light.

Of course the first rockets will be passengerless, for until the nature of the upper atmosphere and outer space is learned, all observations will be made solely by instruments. But granting that there are no destructive rays in outer space and the terrific cold (if such exists) can be overcome, there seems no reason in years to come why a group of scientists could not rise above our atmosphere, say, 500 miles and at their comparative leisure make a study of the earth as seen from above, and of immensities of outer space.

For the present, the aims of the scientific group that will surround the new rocket experiments will be confined to the immediate locale of the earth's surface — that is to say, twenty-five to fifty miles above the surface. Even in this narrow belt there is much to be learned. The science of aviation, still in its infancy, will not mature until the nature of its medium, the air, is fully understood. The laws that govern air currents, the changing temperature and pressure of the upper atmosphere and their effect on the air levels normally traversed by planes can now be learned.

Then there is the eternal question raised by the hypothetical Kennelly-Heavyside layer. When it was found by radio experimenters that radio waves traveling out toward the void apparently were being reflected back to earth, there came into existence the idea of a layer of ionized air which acted as a mirror and prohibited the passage of the waves through it. The mirror was found to be constantly shifting in position and density, and upon its position and apparent activity much of the success of radio broadcasting depends. Now the newest of our means of communication and entertainment, the radio, will have a means opened by the rocket for the solution of one of its most vexing problems. Actual research into the nature and existence of the hypothetical layer can be made by lifting radio instruments above it and attempting to communicate through it with the earth.

Dr. Goddard is working on liquid fuels, which when combined and exploded in a combustion chamber, will expand to many times their own volume (just as in the combustion chamber of an automobile) and by the expulsion of these gases through the rear of the rocket the reaction or recoil will push the rocket ahead. To exceed the present altitude records means to use a greater quantity or more powerful fuels. To use more fuels means increasing the weight of the rocket, and thereby attempting to lift ones self by ones bootstraps. More powerful fuels are, on the whole, those which are so unstable chemically that they cannot be controlled in action. The mixture of liquid hydrogen and oxygen is at present yielding the most promising results.

With the serious problem of the perfection of the rocket to be settled, and with rewards enough in exploration of the neighborhood of our atmosphere to content them, the Goddard committee is not attempting to prophesy the ultimate development of the rocket. The recent statements of Colonel Charles A. Lindbergh, however, that future long distance flying will be made at high altitude, and the statements of German experimenters that long distance flights can be made at high altitude by rocket-driven planes at what seem phenomenal speeds lend a reasonable promise of a new and revolutionary method of transportation. In one of his guesses the late Max Valier, a leading German experimenter, predicted flights between Berlin and New York in one hour. For as he said, a rocket plane ascending to an altitude of thirty or forty miles would have virtually no air resistance and could acquire speeds approaching three thousand miles an hour. As to the truth of this, of course, only the future can tell. But meanwhile rocket travel by plane as well as the remoter vision of travel to other planets, has indeed been given a new life and impetus by the Guggenheim award.

Proposed Constitution of the International Interplanetary Commission

as prepared by David Lasser,
President American Interplanetary Society.

1. Name - The name of this organization shall be the International Interplanetary Commission and its principal office shall be in Berlin, Germany.

2. Purpose - The purpose of this Commission shall be:

(a) The free dissemination among members of all news of the progress and development of rockets, and the various phases of astronautics, including research and experimentation on the nature and physical characteristics of the atmosphere, interplanetary space and solar planets - in brief - of every phase of the astronautical problem.

(b) The planning of the solution of the rocket and astronautical problems by interchange of views among members regarding the best methods of procedure; and the allocation of the several problems to the various members for solution.

(c) The full and complete exchange among the members of experimental and research data on the rocket and other astronautical problems, as experiments progress, in order that comparative results may be studied and the solutions of the problems proceed with the minimum of duplication.

(d) The stimulation of interest in astronautics, and the organization of member societies, in those countries where none now exist.

(e) The sending out of accurate news releases to the world press on the general progress of astronautics, the plans of the member societies and its experimenters.

(f) The dissemination among the members of all books, articles, and other published material bearing upon rocketry and astronautics.

(g) The accumulation of an International Fund for distribution among members, for the carrying on of experiments approved by the Commission.

(h) The accumulation of an International library of books, articles, and other published material on rocketry and astronautics, and the compilation of the more important experimental results of its members.

(i) And such other purposes as the members may later provide, in order that the solution of the rocket and astronautical problems may proceed most swiftly with the maximum of international cooperation toward success.

3. Members - The members of the Commission shall be the several national societies, organized for the promotion of rocketry and astronautics, and individual rocket experimenters subscribing to this constitution; and those that may later be elected to membership by the Commission. Election of new members shall be by a two thirds vote of the existing membership.

4. Officers - The officers of the Commission shall be a secretary, and a Board of National Commissioners, one to be elected by each member society and to include the individual experimenters.

5. Election of officers - The national commissioners shall be chosen by each national society yearly and their names forwarded to the incumbent secretary. The new secretary shall be elected yearly by a majority vote of the Commissioners and individually after the yearly election of the Commissioners.

6. Duties of officers - It shall be the duty of the Secretary to keep the records of the business of the Commission; to send out to members all experimental data, news reports and printed material for dissemination as are set forth in the purposes; to send out releases to newspapers of the work of the Commission and the International progress of rocketry and astronautics; to collect the dues from members and disburse them as may be directed by the Commission; to translate or have translated such material for dissemination to the members as the Constitution provides; to communicate with and encourage the formation of national societies in countries where none now exist; to establish and maintain the international headquarters of the Commission; to keep safe and distribute such funds as the Commission may accumulate, as directed by the Commission.

It shall be the duty of the Commissioners to act as the liaison, between the International Commission and their Societies, to represent their societies at all meetings and deliberations of the Commission, to act for it when directed by their societies in the formation of the policy of the Commission; to collect from the Societies such moneys and experimental data, news, printed material etc. for transmission to the Secretary as are necessary for the fulfillment of the purposes of the Commission; to provide for the safekeeping by bond or otherwise and the expenditure of the funds of the Commission by the Secretary; to audit the accounts of the Secretary; and such other duties as may become a delegate to the International Commission as shall best promote its work.

7. Dues - The dues of each member shall be 100 gold marks per year payable semi-annually in advance at, Berlin, Germany except for individual experimenters for whom the dues shall be 75 gold marks per year, payable semi-annually in advance.

8. Amendments - This Constitution shall be amended only by an unanimous vote of the members; proposals for such amendment to originate by any member Society for transmission by its Commissioner to the Secretary. Such proposals shall be forwarded to the several Commissioners for action by his Society. Amendments may be made bi-annually, it being understood that all proposals originating between bi-annual terms shall be returnable approved or denied by the members before the end of that bi-annual term. Notices of the results of the members' vote shall be forwarded to members at the conclusion of the bi-annual term.

9. Language - There shall be no official language for the Commission. All correspondence except printed material shall be addressed to the Secretary in German. Communications from the Secretary to members shall be in the national tongue of the member.

10. Voting - Each Commissioner shall have one vote and only one in the settlement of all business that may come before the Commission, except individual experimenters who shall have one third vote each.

Annual Report to the American Interplanetary Society

Rendered April 13th, 1931
By
David Lasser

Your officers have rendered to you rather complete reports on their activities. It is not necessary to duplicate the ground they have covered, except to tender them for myself a vote of thanks for their cooperation during this first crucial year.

I realize that our jobs were new to all of us - we were attempting not only to organize something, but to organize something radically new and different, and to carry on our work as amateurs. We were all making our living, which consumed the greater part of our time, in other fields.

I believe, in surveying the work of the officers who have just reported that I can find nothing but evidence of devotion to an idea, that we all believe in; and if some did not accomplish all that they had hoped to, I believe it was because of the pressure of other necessary things rather than disinclination. The members too have been cooperating on a scale, and in a manner that I have never encountered in any organization I have been connected with. The assistance given me in the preparations for the meeting of January 27th was beyond anything that I had expected, and I am very grateful to have had the opportunity to do what I could in our work with our fine membership.

But we cannot afford to look backward too much. As Mr. Pendray rightly stated in his report, our second year will be perhaps a most crucial one. We have so solidly established ourselves before the public that we must fulfill its expectations, or be deemed a failure. We must not only hold our present membership, but we must add to it, in great numbers; we must continue to carry on the work of education of the public, and finally we must seriously begin to consider how and, where we can promote the art of astronautics by actual experimentation.

Covering my own work, during the past year, I can say that perhaps I have learned little. I finish the year with approximately the same beliefs with which I began it with regard to the way the Society can best fulfill its aims. I think that the best way to illustrate what I mean is to review the work that has already been done.

When the Society was formed, I was impressed most of all with the general ignorance surrounding the interplanetary problem - not only among scientists and the general public but among the charter members of the Society, myself included.

We felt certain that interplanetary travel was not only an attainable ideal but also that a method of attaining it had already been found in the rocket. But, beyond that, most of us knew little of the problem.

The first job that confronted us therefore, was to extend our own knowledge of the problem, its limitations and possibilities so that we could inform the world about it intelligently as well as decide intelligently what was to be done to realize our goal.

This aim, described in our Constitution, as the "mutual enlightenment" of its members has, I believe, been carried out in an extremely, successful manner by our primary series of reports by members and by our research on the rocket. I am sure that we have all gained enormously by this series of mutual enlightenment classes and we therefore constitute a group whose knowledge of the problem of rocketry is a valuable source of information for any scientist.

By the time this report is rendered, we shall have practically finished our initial research, covered the first lap of our journey. But we must admit that our knowledge, however great, is still very sketchy, and we must still fill out the general picture by precise practical details.

In order to do this I suggest that the officers for the coming year consider two programs, both aimed at the bringing before the Society of specialized reports on the various phases of rocketry and astronautics. The first is that we bring before the Society specialists in the various sciences associated with our aims and have them give us in lectures the benefit of their experience and knowledge. As an example of this, I have already been in communication with the Chief Engineer of the Sperry Gyroscope Company, and with the Linde Air Products Company to arrange for speakers. The problem of the handling of liquid fuels which is within the province of the Linde Company certainly, can be made more illuminating to us than it is, for it constitutes one of the most difficult problems that confront rocket experimenters.

We need speakers on the mechanics of the rocket, on projectile ballistics, on rocket fuels, as well as on the engineering aspects of our problem.

I warn the officers for the coming year, however, that such speakers, especially men of reputation will be hard to get. The attitude of American technicians, that their association with a project such as rocketry will immediately place them in the category of "dreamers" still exists, and I have met it quite often. This attitude certainly does not exist in Europe and the short-sighted practicality of our American men of science is something that the Society must break down.

That however belongs to another phase of our work, which I will mention shortly. But it is sufficient to say at this time, that it is going to be quite difficult to persuade these men that they're not coupling their names with a group of wild-eyed theorists.

The second way to promote our knowledge of the specialized and more intricate problems is by continuing our own research program. It was understood, I believe that our research on the rocket was to continue until we had covered our subject thoroughly.

I am sure we will all admit that so far we have but scratched the surface thinly. Each of us has an insight now into a specialty. I suggest that the officers for the coming year arrange for the continuation of the research on the lines already laid down, this time getting closer to the heart of our problems. That will certainly be necessary if we are to turn out in, let us say another year, a report which we will be proud of.

I come now to the second phase of our activities. This is described in the Constitution as the "promotion of interest in interplanetary expeditions and travel, and the dissemination of information bearing on the subject."

As your vice-president indicated in his report, when the Society was formed a year ago, little was known in America about the interplanetary question except that a few "nuts" were interested in it. The average man on the street, the members of my own family in fact, merely smiled superiorly when the question was mentioned, and some said to me, "Do you really, seriously think of traveling to the moon?"

Serious scientists on the other hand, used more or less, to wild projects paid the subject no attention whatever. But they were none the less ignorant of the fundamentals of the problem, unless they happened to be astronomers.

Our job, therefore, is to change the attitude of a nation on a project more immense than any that man has attempted: surely a gigantic task. Naturally in one year, we have not by any means succeeded in this, but signs are quite evident that we have made a good beginning, we have made inroads into the general reluctance to accept new ideas. Now we must pursue our advantage with all the energy we have.

The problem is two-fold. One concerns the general lay public and the other concerns men of science. With regard to the first we have already made good progress. Thanks to the attention created by the showing of "By Rocket to the Moon" in New York, the picture is to be sent throughout the country where people will see visually the nature of the problem and how it can be accomplished.

I do not anticipate much difficulty in convincing the general public that interplanetary travel is possible, for in fact their gullibility is often as great as their original incredulity. They can be convinced that the general idea is possible, but often remain unaware of the difficulties, and expect that the project must be realized immediately.

What we need, therefore, is an expansion of the Bulletin - not only in number - but in content of the type and diversity of material. The Bulletin is now in the files of twenty-five metropolitan libraries, it should be in seventy-five. Its size now is eight pages, it should be twenty-four pages with interesting diagrams and general as well as technical articles on the problem and how it is being solved.

In addition, we must ceaselessly hammer at the magazine and newspaper reading public with articles on every phase of our problem. These articles should not only be explanatory of the nature of interplanetary travel but must also enlighten on the way that it can be accomplished. It must show at once that the ideal can be realized, but that the realization is attended with enormous difficulties. In that way only can we create an enlightened public attitude on the question which is the only attitude we should want.

I believe that the members of the Society have been too hesitant about writing articles for the general press. In our research we have uncovered innumerable subjects for the general and magazine press, and so far as I know, with the exception of two men, none of our members have taken advantage of this chance to spread our word and incidentally increase their own incomes.

I urge you to think this over seriously, and I hope that the coming year will see the publication of twenty-five articles as compared to the three or four that have been published by the members thus far.

There exists also need for a good book in English that shall cover the whole subject thoroughly - one that can be used for not only popular exposition but also as a source book for students of science who wish to see the general possibilities in the interplanetary question. I have completed such a book, which is by no means well done and has not yet been accepted for publication; and it should be no bar to any of the other members who feels that he can do a good job of it. *(Italics added. Ed.)*

The conversion of scientists and technical men however constitutes a more difficult and laborious process. The attitude of such men at the present time may be illustrated by a letter to me from a prominent member telling of a conversation with a noted American astronomer. Our member asked the astronomer if he thought that Dr. Goddard would soon accomplish his aim of reaching the moon. According to the letter the astronomer was practically speechless with surprise, he could not seem to understand why an apparently educated, intelligent man of science could put to him such a preposterous question!

Now, we know that such an attitude as this should not exist and we must be prepared to break it down slowly. That can be done in three ways - first by our meetings and our Bulletin, by means of which men of science can see the ways in which the wild dream can be made a reality; and second by personal contact. We must literally make converts, if not members of all the technical men that we meet; so that these men will be centers for the spreading of our propaganda into the technical and scientific magazines, among others. Third, we must carry over our propaganda explaining the various phases of the interplanetary question and indicating where men of science can find it in keeping with their dignity and reputations to associate themselves with our project in some way. As an example of this, I have suggested to one of our members the writing of an article on "The Possibilities for Chemists in Rocket Research." I hope soon to see such an article in print in one of the chemical magazines. There is no reason why similar articles cannot be written on the hundred and one other phases of our problem.

I believe that we are making inroads into the reluctance of men of science to consider interplanetary travel as feasible. Our attitude should be that the Society is really a center of propaganda for an idea; for until the idea has been firmly established no scientist will devote his talents to rocket experiments. And unless we feel the crusading spirit with regard to interplanetary travel, and carry everywhere the conviction that we believe in this thing, we cannot hope that others will flock to our standard.

The Bulletin that we now have is but a crude beginning. There is no reason, given the proper funds, why we cannot publish a magazine of respectable size that people will want to read eagerly and discuss. The material for it already exists, the talent for editing it we already have. What we need is the membership and the funds to warrant its publication.

The obtaining of funds from other sources than our membership is something that I believe that we should not count on. Unless we can maintain our organization as a self-supporting entity, we cannot hope for the respect from people that must come before a monetary contribution.

Your vice-president has already covered quite well the membership situation. It is improving perceptibly, though I am frankly disappointed. However I believe that much of the lack of enthusiasm to contribute money to membership can undoubtedly be laid to unemployment and the business depression. With 6,000,000 people out of work and perhaps ten percent of the population deprived of its income, people are undoubtedly reluctant to spend money for other than immediate necessities. I feel convinced that when the general industrial situation clears up that our membership will grow rapidly. But in order to have the gains that we need, I can see no other way than through the efforts of the members we now have.

To me, there can be nothing lukewarm about the interplanetary question. I cannot, conceive of a man or woman who feels simply an indolent interest in it. I cannot see, therefore, why our members - active and associate - both should not feel that crusading spirit and constant desire to inform friends and relatives of the possibilities inherent in our work.

The growth of the Society would be amply expanded during the next year if our present members were to make the resolution to bring into the Society ten members each during the coming year. For if we do not feel proud of our efforts, and do not talk about them, who will? I ask you therefore, on behalf of your own vision of the future of our work to go out of this room determined to bring into our ranks ten people each before the day for the next annual meeting arrives. It will mean not only something to them, but something more to you, for the Society can return to you in its Bulletin and its meetings and its experimentation a dividend for your efforts.

With regard to the active membership itself, you all know that we need several good astronomers, physicists, and practical engineers to mention but a few. I don't doubt but that here in New York there are the very men that we want, and that they want us. It means only some connecting link to bring us together. You men and women are that link.

With regard to the expansion of our active membership outside of New York, I believe our attitude is quite clear. We do want to make the Society truly national, and to have ultimately little local groups where men and women can gather together to discuss the fascinating problems that surround the interplanetary question. In that way only can the spirit of our enterprise really permeate the nation. But I believe that such decentralization now, that must accompany the creation of branches, would be fatal, I therefore recommend that we go out vigorously after out-of-town members, with the intention of establishing branches by the time that the next annual meeting comes around.

I come now to what is perhaps the practical end of our work, described in the Constitution as "the stimulation by the expenditure of funds and otherwise of American scientists toward a solution of the problems that bar our way to space."

Directly, the net result of our first year along this line may be characterized as precisely zero. Yet I am not alarmed at this. I am aware, for example, that a good many members want the Society itself to go in for some immediate experiments. In general I am in sympathy with that attitude but

I believe that with the present state of our treasury and our knowledge of our subject it would not be wise. We must remember that we are in this thing for a long time, at least until there is no longer any need for us.

Before we can do any experimenting we must do some planning so that not only will the experiment mean something to us but also it will show the world that we are approaching our problem intelligently as well as enthusiastically.

Our lack of activity during the present year along the lines of experimentation was due simply to the fact that there was but one call for assistance and that assistance we were unable to give.

If the Society means anything, it will be called upon increasingly in the future by experimenters for help of one kind or another, and we must have some machinery to meet those calls. For that purpose I respectfully suggest the creation of a Committee on Experiments whose job it will be 1. To decide where inexpensive but purposeful experiments can be made. 2. To lay out a general program for a series of experiments with costs 3. To cooperate with research workers who wish to do experiments of their own and who solicit the Society's advice 4. To advise the Society on how it may experiment to advantage on its own. This Committee can, I believe, be one of the most useful that the Society has, and I suggest that it be composed of men who have a personal flair for such work, and a practical sense of how it can be done. I believe that an Amendment to the Constitution must be made to provide for this Committee as I feel it should be as permanent a Committee as the Membership, nominations and Executive Committee. I suggest that the Society consider it, at this very meeting.

We must certainly push vigorously the experimental side of our work. And we can do it best by stimulating as many individuals and groups to carry on experiments as possible. If the Society, by its meetings, its Bulletins, and its general educational material can provide the incentive to ten or twenty men of ability to begin rocket experimentation, certainly we are getting ahead faster than if the Society devotes all of its own energy to experiments. I firmly believe that if we can study this question and point out just what can be done, that a crop of experimenters will arise over night. That is the way America does things. And out of this hasty enthusiasm will come a number of contributions of real merit. With the utmost respect to Dr. Goddard, I believe that America needs more of his caliber. It is up to us to find them.

Another study that I believe should be begun immediately by the Executive Committee is what I would call a war campaign for the carrying out of our aims. I recommend that the Executive Committee lay out a program how it would use a given sum of money were it to be contributed to the Society or to be administered to it. How in short could we show a man or woman or organization of means that $5000, $10,000 or $50,000 could be spent advantageously in promoting astronautics.

This study should be made carefully, it should be completed and put away for future use; subject to revision as conditions dictate.

Another project that I recommend the Society bring to completion is the formation of an International Interplanetary Commission. I have already opened the way for its formation, and I believe that if it can be successfully accomplished it will do much to promote the fulfillment of our aims.

Such a commission would, in my opinion establish international headquarters which will be a center for world information on all phases of the interplanetary question, as well as a center for the formation of national societies in countries where they do not at present exist.

There should be an international press service where news on world developments will be gathered and translated into the various languages and disseminated to the various national societies. The International Commission would also cooperate in the actual fulfillment of our goals to the fullest extent possible. I believe that much can be gained by the international exchange of experiences not only on the ways to arouse national sentiment for our aims but also of experiences in experimental work.

I anticipate, naturally that in dealing with a question such as the rocket which may be a weapon in future warfare, that full cooperation will not be possible, but a beginning can be made. Such international cooperation should be begun and pursued energetically, for it seems to me that the solution of the interplanetary problem is too large to be localized in any group or nation. I can foresee the building of the first space ship only as a joint effort of an united earth.

M. Esnault-Pelterie and Professor Rynin of Russia already indicated their agreement to this project. Mr. Pendray is approaching the German group to determine what can be done with them. But I mention the formation of an International Commission as one of the projects that the Society should consider as a means toward its goal.

I recommend also that the Committee appointed to effect the incorporation of the Society carry it through at the first practicable moment. The Society has already discussed the matter fully, and I can state more emphatically than ever that I believe early incorporation to be vitally necessary if we are to grow and expand.

In conclusion, I can say that aside from the slow growth in membership I am quite satisfied with the results of the Society's first year. I felt at the beginning that if all we could do during that first year was to establish firmly in the nation's mind that there actually was an organization of serious-minded men and women whose purpose it was to make interplanetary travel a reality, then we had accomplished enough. This, I think you will agree, we have already done.

I believe that if during the coming year we will energetically pursue the increasing of our membership, the broadening of our Research program; the extension of the Bulletin; the cooperation with experimenters; the formation of an International group and the sending out of articles for general and technical publications that we will have completed a good year.

I think that despite the enormous difficulties that must be leveled before a successful space flight can be made, that it will be accomplished within the lifetime of all of us. We should all feel a great enthusiasm for our work in helping to make it a dream come true. We have but started and the difficult tasks lie still ahead of us. I want to thank you for the honor of being the Society's first president and being permitted to do what I could. It was a pleasure, as much as it was a welcome task.

The Rocket and the Next War

by David Lasser*

*President The American Interplanetary Society 1930-32 - Author "The Conquest of Space" Penguin Press, N.Y. 1931

During the past few years the press of the world has carried many stories on the development of the rocket as an instrument of science that could extend mightily our conquest of space.

It happens, however, that the qualities of the rocket that make it a revolutionary agent of transportation, also make it a powerful instrument of destruction. And so great are its potentialities, that if energetically developed, the rocket may play a vital role in future conflicts. Military strategy may require an instrument which is now possible only by the rocket.

As we look back to the four bloody years of the Great War, during which two armies of millions of men struggled against each other, we realize that the old days of war are over, and a new era has begun. It is appreciated now that the conflict of millions of men for a decision in the field, is a long costly process both for the victor and the defeated. I call your attention to the last great German drive in 1918, in which Germany threw everything she had into an attempt to crush the Allies on the Western front. Before the 1918 drive started, Generals Hindenburg and Ludendorff speaking before the German Reichstag stated that they could beat the Allies but that Germany must be prepared to suffer 1,504,000 men in casualties during the year. Before the war was ended, their losses had mounted to that figure. But despite this enormous sacrifice Germany was beaten.

I believe that military men will no longer be permitted to squander their nation's manpower and riches in mass conflicts of millions of men. Instead, they will try to paralyze their enemy by cutting off the supplies and food of its army, and destroying the morale of its people.

It is probable that future conflicts will see all factories, all railways, all industry harnessed to the war machine. The term non-combatant will hardly exist. Shrewd directors of military affairs will realize that for victory they must first cut off the life blood to the enemy army colossus, before they can destroy it. If they cut off the communications and transportation of the enemy army, bomb munitions centers out of existence, raid enemy centers of population so effectively that the enemy's morale will collapse, the war will be over quickly.

That strategy was in the minds of military men during the closing days of the last war, even when the Allies delivered smashing blows at an exhausted Germany, they saw the futility of beating the German army to its knees. For if they had attempted to, there would probably have been nothing left of either army. Instead they hammered at the railway lines of the Germans, and the success of it may be judged by the statement of one of the German emissaries who signed the armistice.

"We do not regard ourselves as defeated", he said, "but since our main railway line has been broken almost continuously by long range shell fire for two weeks, a half million of our men are starving, having had nothing to eat for the past four days. We must therefore cease."

Two military instruments able to destroy and strike terror into the heart of an enemy are possible in the present state of science - the air raid and the long distance bombardment. The rocket is the principle that must and will be used to make both the air raid and the long distance bombardment terribly effective. And because they will be so successful, the brunt of future wars will not fall upon the armies in the field but upon the nations behind them, in the cities, the factories and on the railways.

Consider first the long range bombardment. Here on the map of Europe we have two nations at war, their armies already at the frontiers. Adhering to the new strategy, each will attempt to spread terror and destruction in the heart of the enemy's country.

The guns that are used must first have a range sufficient to send shells into the heart of the enemy country; they must be accurate; and third they must be mobile, to operate efficiently and escape counter-bombardment. For these requirements present types of artillery fail and a new type of artillery must be developed.

The best example of the weaknesses of present artillery is found in the German guns that bombarded Paris from 75 miles during the great war. They were the most powerful pieces of ordnance ever built, but as military weapons they were failures.

In order to send a shell 75 miles from one of the Paris guns it was necessary to give it a muzzle velocity of 5500 feet per second. At one moment the shell was resting in the breech, and 1/50 of a second later it was flashing from the muzzle at 3700 miles an hour. The power to provide this great energy had to be applied instantaneously and it would have been sufficient to lift from the ground five ships the size of the Leviathan.

For the operation of such a shell a tremendously heavy and powerful piece of ordnance was necessary; and heavy and complicated auxiliary equipment to support it. The Paris gun, I speak now of the muzzle, was actually about 130 feet long, it required a base made of concrete 35 feet square and 15 feet deep, the whole unit weighing hundreds of tons.

Obviously the first disadvantage of such guns, and this goes too, for all similar artillery, was that they lacked mobility. Days were required to emplace and position a gun, and days were required to remove one. Not only that, but also their positions could be located by opposing batteries. The Germans did try to smother the terrific detonation of their Big Berthas by opening a general bombardment along that section of the front at the time of firing. Despite that, not two days after the first shells fell on Paris French batteries had located the site of the guns and shelled them, wrecking one gun and killing the crew of another. Since modern airplane observation of artillery positions, and sound detectors are becoming perfected, the location of big gun positions becomes easier and easier, and the building of larger guns means only the invitation to destruction.

Furthermore, because the guns are fixed, they are liable to capture should the tides of battle suddenly change. For their maximum effectiveness in shelling an enemy country, they must be located close to the front. If the enemy were to break the line the gun crews would be unable to remove the guns, and possibly not even have time to sufficiently wreck them to make them useless.

These disadvantages that I have mentioned thus far, are not only inherent in present artillery, but will become more and more serious as the attempts are made to build guns of longer and longer range.

The 75-mile gun not only lacked mobility, but also accuracy. It was discovered that the tremendous rush of highly compressed and heated gases up the muzzle at each firing, not only eroded the muzzle but heated and weakened the gun. The pressures developed within the muzzle were enormous, and no metal could withstand them very long. Fifty to 100 shots were all that could be obtained from these costly monsters before they were useless and had to be returned to the factory for reboring. And because of the erosion and wearing of the guns, the accuracy suffered badly. A gun aiming at its 75-mile mark often overshot or undershot by nearly 5 miles. Furthermore because of the complication in the firing of a shell, the tremendous strain imposed on the gun and the elaborate calculations necessary, one shot in 20 minutes was the best that gun could give.

Then consider the limits of the range of present types of artillery. Because the shell is emitted from the gun at its maximum velocity that velocity must be sufficient to give the shell its ultimate range. But the more the initial velocity is increased, the more the air resistance increases, the resistance increasing much faster than the velocity. It will possibly be true that even if immensely heavier guns are built to withstand the tremendous strains imposed upon them, and even if somehow these giants could be made portable, it is doubtful if their range could be increased greatly. An increased muzzle velocity would be eaten up quickly by enormously increased air resistance; the added power put behind the shell serving only to heat it by its rapid passage through the air.

The builders of the 75 mile gun realized this. Their only hope of attaining even that range lay in shooting the shell so high into the air that it would pass into the rarefied levels during its flight, where the resistance would be reduced. And, in fact, about three quarters of the flight of these projectiles were through a virtual vacuum between 12 to 24 miles above the earth. Had the 75-mile gun been fired in a total vacuum its range would have been not 75 miles but 160 miles - more than twice as much. I mention this to illustrate the effect of the enormous air resistance encountered in the lower levels of the air.

What is wrong about this method of propulsion is that the maximum air resistance is encountered in the lower, denser levels of the air where the velocity is greatest. All factors thus conspire to kill the shell's kinetic energy and reduce its range.

I hope that this brief summary of the Paris gun illustrates what I mean when I say that modern artillery is by nature not equipped to fight the future war of long distance bombardment.

What is necessary is a form of shell that leaves the gun at a small velocity and increases in speed as it ascends through the air and only reaches its maximum speed where the air is rarefied and the air resistance is small. The velocity curve of such a shell would be the opposite of that of present artillery shells. Secondly, the shell must have a possible range considerably in excess of all other forms of artillery, in order to be able to outduel it, and to strike at enemy centers located far inland. And thirdly, its equipment must be simple enough to permit rapid firing and mobile enough to be shifted quickly to wherever it is needed the most; and be moved when it is necessary to escape counter-shelling or bombing from enemy planes.

Fulfilling these conditions perfectly we find the rocket method of propulsion; and because military men are now alive to the possibilities of scientific instruments I have no doubt but that before the next great war the possibilities will be realized and great advances made on it as a means of destruction.

In order to make clear how the rocket satisfied the peculiar needs for artillery I will briefly outline its principle of operation.

The rocket operates, in a manner similar to a cartridge. The explosion of fuel in a confined space with an opening causes a tremendous pressure of the resulting gases. They press against all the walls of their combustion chamber, being expelled to the outside, and pushing the rocket ahead.

The principle is a very old and honorable one, and it is of a compelling simplicity for it needs no complicated equipment for its operation.

To use such a rocket as a shell, we need only add a nose which may be filled with high explosives, shrapnel, poison gases, infectious germs, in fact, anything that civilized man can conceive of, to kill his nations enemies.

We have, as a means of propulsion, instead of the single tremendous thrust of artillery, a steady powerful rush created by the burning of the fuel. This offers the great advantage that no heavy complicated equipment is necessary to shoot the gun, for there is no enormous recoil to be taken up.

The equipment necessary to operate the rocket shell would be absurdly simple. Instead of tremendous ordnance and fixed gun emplacement, there would be necessary only a light gun to give the shell its direction and the necessary rotation for stability. Such equipment could be very mobile and weigh a small fraction of what present artillery weighs.

The shell, projected from its gun would be propelled by the burning of the fuel in the combustion chamber, until the fuel had been exhausted. It would then continue on by its own momentum, until it reached the limit of its upward flight, and would then descend with terrible speed to strike with fearful force whatever it hit.

Instead of a few guns, which like the Paris gun, could be fired only a few times a day, a future army could have hundreds of rocket batteries delivering an incessant bombardment to destroy whatever they were trained upon.

The Germans admitted that their intermittent shelling of Paris allowed the populace not only plenty of time to carry on its business and to take over leisurely but also allowed it to recover from the nervous and mental affects of the shelling. But if Paris, the heart of the French and allied governments, had been shelled incessantly, if the railroads entering and leaving the city had been deluged constantly by monstrous shells, it is obvious that the French government might have collapsed and the war ended or a rebellion of a panic stricken populace ensued.

In the matter of range too, the rocket far excels artillery. The rocket does actually start at a small speed, encountering little resistance where the air is dense. Its speed increases, as the air becomes thinner and it reaches its maximum speed near the summit of its flight where the air is rarest. Because a rocket can carry its own fuel, it can be sent over distances utterly impossible with artillery. It can utilize fuels such as oxygen and gasoline that has three or four times the energy of gun powder, and can utilize them to a much greater efficiency. The powder that is used in artillery gives only about 20% of its energy to the propulsion of the shell, while experimenters, notably professor Goddard, with crude equipment, obtained efficiencies of 60% with rockets.

Shells sent two hundred miles are well within the range of possibility with the use of the rocket. What would this mean in terms of warfare? Berlin, Paris and London would practically by within shelling distance of their frontiers. In European warfare each nation could devastate each other's cities by a hail of gigantic shells, from which there could be no relief. Cities, munitions works, all of the creative and destructive facilities of man could be destroyed without the necessity of either opponent leaving its center of activities. Such shells carrying gas could spread terror and death in great cities, and make it virtually necessary that each citizen down to the smallest babe be prepared to put on instantly a gas mask and wear it until the danger was over.

Armies facing each other from hundreds of miles could shell incessantly each other's lines of communication and paralyze all railway traffic, so that starvation and lack of supplies would make the field armies helpless. General Jan Smuts, speaking recently of the next war stated that, "It will fight with new and unheard of chemical and biological weapons. It will cover the fair land and the great cities with poison and disease germs. It will saturate vast areas with a deadly atmosphere."

General Smuts assumed that it would be the aim of a nation in the next war to really attempt the attack on the heart of its army. Commenting on this the New York Times stated editorially that "tests of anti-aircraft guns at Aberdeen may reassure those who have feared that there is no defense against war planes equipped to drop poison gas bombs on civilian populations and destroy cities by powerful explosives. What was to happen in the next war," the Times goes on to say, "has been described in terms to frighten every nation out of its growth, if the predictions were to be taken without a thought of means of protection. In such a vast convulsion the airplane would be the weapon of destruction. If a defense of it from the ground is contrived countries will feel safer."

Now it is obvious that the Times writer did not consider the possibility of one nation bombarding another by long distance shells. The limited range of such shells obviated its ability to spread such destruction in great centers of population. But with the use of the rocket, all that General Smuts has pictured may be possible, and so far as we know there is no protection that a great city can use against a rocket.

As Americans we ask what that would mean to us. In our isolated position can we feel secure against such devastation from a distance?

How could the rocket be effective in naval warfare? The present towers of protection against an enemy invasion is the air force, navy and great coast defenses with sixteen-inch guns throwing shells some 30 miles. Naturally nations, since invading warships cannot carry such heavy guns as a coast defense nation like America have, in the past, felt security. Ships could not come within dueling range of its forts.

But what would be the situation if enemy warships hundreds of miles beyond the coastline were to reduce coast defenses by rocket shell fire. The coast guns would be toy cannon in range compared to the naval rocket batteries. And with their fixed positions the forts would be easy targets. Warships armed with rocket batteries could shell from comparative safety Boston, New York, Philadelphia, Washington, Richmond or the large forts in the Pacific if we had the protection only of our present artillery.

All that the invading fleet need do is hold off our air force, with their own air force and with their anti-craft guns; hold off our navy with their own (and they can do it easily by their ability to outduel us), then their destruction of our coast would follow.

This would be possible since no tremendously heavy armament is necessary to shoot rocket shells. They carry their own propulsion. It is merely necessary to shoot rocket shells, aim them, give them a small start and they propel themselves.

Small vessels, quick moving, eager, agile, hundreds of them could carry these rocket batteries and deluge our coast with a rain of fire and lay waste to cities a hundred miles inland.

Those who doubt the destruction that can be wrought by the rocket need only turn back to the Napoleonic wars, and the experiences of one Sir William Congreve. Congreve was a great believer in the rocket as a weapon of warfare. His enthusiasm was so infectious that he obtained a grant from the British military authorities, took the paper rocket that had been unchanged for centuries, gave it a casing of iron, improved its internal construction and built rockets far larger than had ever been constructed. He filled the rocket heads with lumps of iron and inflammable materials and securing a few small boats anchored several miles off the French town of Boulogne. Out of range of its forts, he sent the rockets off. They burst in the town on landing, exposing a great mass of flaming surface to set fire to anything that they touched.

With such simple equipment, Congreve succeeded in terrorizing the French populace until finally Napoleon ordered his military men to build rockets at any cost. Although the construction of Congreve's days were absurdly crude, it appeared for a time as though they might decide the Napoleonic wars. But they did prove that with ridiculously small equipment, with no ordnance, they could become a terrible engine of destruction.

With a crew of untrained men, Congreve shot 200 rocket projectiles into Boulogne in a half hour, and in 10 minutes the entire town was on fire. In Copenhagen in the next year he repeated his work. With sixteen men he shot 300 shells into the town within a short time again setting it on fire.

His report to the British government on the rocket may perhaps be applicable today. Because in naval warfare it is possible to project from small vessels shells that would ordinarily require large ships, Congreve called rockets, "ammunition without ordnance, the soul of artillery without the body...The British navy", he added, "could bombard French maritime towns so that the war might be made the scourge of a large part of its population; and a discontent and terror, without any palliative would soon find its way into the interior to force the French to end the war."

The rockets used by Congreve, although they played an important part in his day were soon discarded when improvements in rifled artillery came along, and for nearly a century the rocket has not been used as a projectile. But today with the new science of rocketry developed, with a new necessity made for it by military strategy, the rocket for long distance bombardment is likely to add a new page to the science of warfare!

I come now to the second agent of future military strategy, the air raid. That, too, was employed by Germany during the war, as a means only to destroy Allied morale, and it too was a failure. The Germans accomplished only enough to arouse the bombed cities to a new pitch of resentment, rather than to an agony of fear.

In the first place, after the first few raids, whether by plane or Zeppelin, the cities had plenty of warning of the approach of the raiders. Sound detectors and observers were able to report of the coming of the enemy in sufficient time to allow the authorities to throw the cities into darkness and to organize a means of defense.

What is evidently necessary to make a successful air raid on the heart of an enemy country is first complete surprise; second, swiftness of execution. If a form of air raider were developed that combined these qualities, it would become a dreadful menace. It could bomb a great city without mercy.

I believe that this agent is found in the rocket-plane. Imagine America at war with a nation across the seas, and three thousand miles apparently separating it from danger. It is quite possible that one hour after war had been declared, an enemy equipped with rocket planes could be bombing its coastal cities. Before America had recovered from its surprise the enemy raiders could have shot their thunderbolts and swiftly escaped.

This is impossible with present types of airplanes, just as the long distance bombardment is impossible with present forms of artillery.

The present plane is a vehicle of the lower air levels - it cannot operate efficiently higher than seven or eight miles above the earth. In the lower levels it not only encounters the great air resistance that limits its speed, but also it is within the range of visibility and detection by the attacked. The droning of airplane motors, the sight of them approaching, is warning to prepare for defense. Furthermore, airplanes for long distance raiding are always at the mercy of the weather, especially the heavy, slow moving bombers.

The rocket plane however, suffers no such disadvantages. Because the rocket can operate where there is no air, and in fact, achieves its greatest efficiency in a vacuum, rocket planes could ascend in an air raid to a thirty-mile altitude, and in this virtual vacuum achieve a speed of more than three thousand miles an hour. At their great height, where they would be practically invisible, a fleet of planes could span the ocean in a single hour and with their acquired momentum, with motors silent, they could swoop down from the blue upon a great city and deliver their terrible loads of death. Quickly ascending to their thirty-mile level, beyond all pursuit, they could return in safety to their base.

Such planes built by the thousand would constitute an avalanche of death from which there could be no escape. What nation could withstand merciless death from the skies and being shelled continually from the water?

Dr. Goddard, the noted American rocket experimenter has already patented a type of rocket plane capable of travelling 3000 miles an hour. Piccard of Belgium, Nebel and Oberth of Germany, Esnault-Pelterie of France, Darwin Lyons in Italy, are at work on rockets - with the eventual idea of building rocket planes. And there is already evidence that Japan and Russia are taking a lively interest in the possibilities of the rocket.

Such warfare will not have the same savagery and blood lust as distinguished past wars. The operators of rocket batteries will have no physical contact with their enemies. They will work coolly, methodically and scientifically to hurl a ceaseless rain of death and destruction hundreds of miles. The field armies will do little more than hold the frontier, until the enemy nation has been brought to its knees.

Professor Hermann Oberth, the noted Austrian rocket experimenter, speaking before the Vienna Meteorological Institute recently stated that research on the rocket as a weapon of warfare was being carried on vigorously in the War Office of every nation. And that the world must be prepared to face the rocket as actuality in future wars.

Of the truth of his statement, with reference to present military research I have no knowledge. But I do not doubt that Herr Oberth's statement could possibly be true. The fact that no official announcements have been made mean nothing, for some War Departments hide their activities in mists of secrecy in order to perfect a surprise weapon, and others broadcast them lavishly in the hope of frightening their possible enemies.

The lack of information does not presume that nothing has been accomplished. But granting as we must, that the rocket as a threat to civilization will exist, what attitude should be taken toward it? We can either pretend ignorance of such a horrible thing, or we can face the possibilities. I prefer personally to face them. For if military use is to be made of the rocket, it is much better that people should know it, and realize the risks that they take when they go to war. It is much better that the world should know of it. It is possible that just as the Great War hastened immensely the development of aviation, so the use of rockets in a future war may bring closer the day when rockets will annihilate time as a servant of transportation.

I personally feel that if the rocket is ever used to the extent that I have pictured I would rather that the knowledge of it be erased from our minds, and that we lose also its wonderful possibilities in peace. But I do not believe that we must therefore cease work upon the rocket any more than we should cease to develop airplanes. If it is not developed, some other weapon will be; for when the desire to destroy is alive, means will be found by men to perfect the weapon. I would rather that we recognize that the rocket is both a weapon of war and a means to a higher and better civilization. And while we develop it openly for civilization, we should remain continually aware of its destructive properties.

I have the rather naive belief that as soon as wars become as terrible as the rocket can make it, that war will cease, because then, I think, the organized opposition of the earth's people will prohibit them.

May 23rd 1936

Mr. David Lasser,
Workers Alliance of America,
Room 510 Burchell Building,
817 14th St. N.W.
Washington, D.C.

Dear Dave:

I was pleased to see your picture on the front page of the Literary Digest a week or two ago and also glad to hear from you.

Unfortunately the "Conquest of Space" is all exhausted. J.J. Little & Ives destroyed the unbound copies, which was their privilege, in order to keep from piling up more storage. If you have any copies left, I advise you to hold on to them. They probably will be more valuable later since we have several orders and I understand the book has become a collector's item.

I will write to the people whose letters you sent me telling them that the book is out of print.

Sincerely,

G. Edward Pendray

Arthur C Clarke, CBE

COLOMBO 7, SRI LANKA.

Telephone: Celltel: Fax: Cable:

March 1992

Dear David -

Heartiest congratulations on you're 90th! I hope this reaches you in time...

I do hope you're keeping well. I've just sent the photos of our meeting to my biographer.

Another old-timer in today's mail - Ray Gallun!

All my best wishes -

April 12, 1992

Dear Arthur:

My many thanks for your fine note. Unfortunately it got mixed up in the letters I received and only after desperate, ceaseless searching did I recover it to get your address. Your letter brings back many memories (wasn't it sixty years) ago as venturesom youngsters daring to ignore the savage criticism thrown at us.

As you can see from the attached, the party was a real success.

Since my doctor told me that if I behaved myself, I had a real chance to have another such party at my 100th birthday, my chief aim is to finish a book on the origin of the universe which I have been working on for five years.

I admire your achievements tremendously and perhaps if you ever touch base in the western part of the USA, we might have a reunion.

I don't know whether you remember this but we amalgamated our little organization with another one that became what is now the renowned American Institute of Aeronautics and Astronautics. As a gesture of extreme generosity they still consider me as the "founding President." I am invited to their annual convention later this month. But, unfortunately my health will not permit my making the trip.

Since I do not know where on earth you are now, I am risking sending this to your Colombo resting place.

I fervently hope that your good health continues and you have many more adventures facing you. With warm best wishes,